The Oil
Muddle

The Oil Muddle: Control vs. Competition

James B. Ramsey

Ethics and Public Policy Center
Washington, D.C.

338.2728
R180

Library of Congress Cataloging in Publication Data
Ramsey, James Bernard.
  The oil muddle, control vs. competition.
  Bibliography: p.
  1. Petroleum industry and trade—United States.
2. Petroleum law and legislation—United States.
3. Competition—United States. I. Title.
HD9566.R35   338.2′728′0973   81-15251
ISBN 0-89633-049-4   AACR2

$6.00

Contents

Appendixes

Foreword

"BIG OIL"—MORE PRECISELY KNOWN AS the worldwide petroleum industry—has always been newsworthy and controversial. It was big news in 1971, when the OPEC cartel started to flex its muscle. It continues to be big news in 1981: the day these words are being written, August 22, two front-page stories in the *Washington Post* underscore the continuing importance of petroleum in domestic and international affairs—"Saudis Freeze Price Until '83: Oil Production to be Trimmed"; "U.S. to Buy Mexican Oil for Strategic Reserve." And big oil will still be in the news in 1991 and 2001.

The endless debate over oil has often been long on emotion, short on facts, and bedeviled by hoary myths about conspiracy, inordinate power, and the badness of bigness. The level of public discussion scraped bottom in 1979, when the long lines at the gasoline pumps led many impatient motorists, journalists, and politicians to charge that the "gas shortage" was caused by collusion or conspiracy among the Seven Sisters. Later that year President Jimmy Carter accused the oil companies of reaping "windfall profits," though their average return on investment was about the same as that of heavy industry generally.

The conspiracy theory was knocked into a cocked hat a year later when two carefully researched government reports, one by the Department of Justice and the other by the Department of Energy, concluded that the long lines were caused not by a real or contrived shortage of gasoline but by a maze of complex, confusing, and unhelpful federal regulations. Big government, not big oil, was to blame.

In the present volume, Professor James Ramsey, chairman of the economics department at New York University, addresses the continuing oil crisis in clear, concise, and non-technical language.

He presents his facts about the petroleum industry within the larger framework of the U.S. economy, dealing with central economic concepts like the market, profit, and competition.

Some surprising facts emerge when we take the long view of the oil industry. For example, Department of Energy figures show that from 1941 to 1980 the average *real* price of a gallon of leaded regular gasoline increased only about two cents. The average 1941 price was about $.19; the average 1980 price of $1.19 would be about $.21 in 1941 dollars.

Professor Ramsey concludes that market forces are more effective than government regulation in curbing the power of OPEC, maintaining an adequate supply of oil for civilian and military use, and keeping oil prices at a realistic level. "We do not, never did, and will not in the foreseeable future," he says, "need a Department of Energy to supplant the normal workings of the market." If he is right, perhaps the unceremonious interment of Adam Smith by the Marxists and Keynesians was premature.

Professor Ramsey is the author of many articles on economic and econometric theory and of several books, including *Economic Forecasting—Models or Markets* and *Bidding and Oil Leases*. Before going to New York University he taught at Michigan State and the University of Birmingham, England, and was a research fellow at the Hoover Institution.

This study is a product of the Ethics and Public Policy Center's Business and Society Project. We acknowledge with gratitude the permission to reprint tables and graphs prepared by others in the appendixes of this book. We also express appreciation to the persons who served as critical reviewers of the manuscript: Joy Dunkerley and Milton Russell of Resources for the Future, Martin Herz of Georgetown University, and Stephen Mayerhofer of the Center staff. Carol Friedley Griffith edited the study. As in all Center publications, the author alone is responsible for the facts selected and the conclusions reached.

ERNEST W. LEFEVER, President
Ethics and Public Policy Center

Washington, D.C.
August 22, 1981

Drawing by Stevenson; © 1979 The New Yorker Magazine, Inc.

Introduction

The facts, ma'am, just give me the facts.
JACK WEBB IN 'DRAGNET'

UNLIKE THE TV CRIME SLEUTH, anyone who wants to find out what is going on in the oil business must ask for more than just "the facts." The facts require interpretation. They must be seen in context, with some knowledge of the overall economic system.

The following headlines and news clips serve as a partial summary of the "facts" about the oil crisis with which we have been bombarded during the past eight years:

Independent Gasoline Dealers Ask Legislature to Assure Supplies Before They're Forced Out [*New York Times*, May 16, 1973].

FTC Study Links Shortage of Gas to Profit Motive [*New York Times*, July 8, 1973].

Price Conspiracy by Eight Oil Concerns Charged by FTC [*New York Times*, July 18, 1973].

Oil Profits Under Fire [*Time*, February 4, 1974].

More Profit and Suspicion [*Time*, May 6, 1974].

Study Urges Check on Oil Companies to Guard National Interest [*New York Times*, February 11, 1975].

Inside the Big Oil Game: Playing With Billions, Shuffling the Taxes, Gambling on Discoveries [*Time*, May 7, 1979].

Playing Politics With Gas [*Time*, May 28, 1979].

Refiners Are Suspected of Using Heating Oil to Warm Their Profits [*Wall Street Journal*, September 14, 1979].

Citizen Groups Say Oil Profits Are Understated [*Washington Post*, November 15, 1979].

It [Robert Sherrill's article "The Case Against the Oil Companies," *New York Times Magazine*, October 14, 1979] confirms my suspicions that oil companies control the availability of

1

oil, manipulate its price, and reap unjustified profits. . . . How many will have to suffer economic troubles and physical ills, how many will have to die before the citizens of this country yell loudly enough to make government officials use the means at their disposal to force the oil companies to act responsibly and make oil available to everyone at a truly fair price? [the Reverend R. C. Block, Berkeley Heights, New Jersey, as quoted in the *New York Times*, November 18, 1979].

The embarrassment of the oil industry's 1979 record-breaking profits . . . the overflowing abundance of cash-flow which generates widespread oil company diversification—not with stockholders' but with consumers' money—into such fields as electric engines, department stores, grocery chains, almond orchards, newspapers, and medical equipment . . .

If decontrol, higher prices, and higher profits are required because the oil companies simply refuse to drill for oil now, when they are already earning record-breaking profits and diversifying widely, then perhaps we should do more than merely lament the plight of the poor [Joel R. Jacobson, commissioner, New Jersey Department of Energy, as quoted in the *New York Times*, November 25, 1979].

That is part of what we think we know about the oil crisis. What most Americans really want to know, however, is: Why have the prices of gasoline and heating oil gone up so much? Will they continue to rise? Why did we have to wait in gasoline lines a few years ago? The system seemed to be working well before—what went wrong? Will we have enough oil to heat our homes this winter? Why can't the government control the oil companies?

My intention in writing this book is to provide a nontechnical economic analysis of some of the history of the domestic oil industry that will contribute to an understanding of the current oil situation and help to indicate which policies are feasible. In a sense this book is as much about *perceptions* of the oil industry, its critics, the role of government, and markets as it is about anything else. It will have been successful if the reader's media-sponsored impressions are subjected to the discipline of an economic viewpoint and if he becomes more able to distinguish facile and misleading arguments from economically valid ones.

There is a tendency in recent writing on the oil industry to talk only about the last decade. We need a much longer perspective in order to perceive why we are where we are. Furthermore, most of the books designed for the nontechnically oriented reader do not interpret the facts through the insights provided by economic theory or, at best, use only the simplest of economic notions. Economic theory provides the sieve through which we can separate the relevant facts and plausible arguments from the irrelevant and implausible, the economic wheat from the rhetorical chaff. That economic understanding is as important as knowledge of the facts is illustrated by a conclusion in the previously mentioned article by Robert Sherrill entitled "The Case Against the Oil Companies":

> But then, in 1930, independent oil operators began bringing in the fabulous East Texas field. Here came the legendary Spindletop and subsequent discoveries, some with wells gushing 10,000 barrels daily. The major oil companies had had little to do with the discovery of the East Texas fields. So, the majors, pretending to be horrified at the wildcatters' wasteful production methods, demanded that the government of Texas force the small companies to "conserve"—i.e., sell only as much as the big companies said they should sell. When the small companies refused to knuckle under, the majors gave them a rugged dose of competition by slashing the price they would pay for crude— from nearly $1 a barrel to 10 cents [*New York Times Magazine*, October 14, 1979, p. 101].

The clear implication, one that is sustained by the rest of the article, is that the "majors" (the largest companies) acted as a cartel in defiance of the antitrust laws and the Department of Justice. They "controlled" the market for oil in the country and lowered the price in order to drive out small operators who were threatening them with genuine competition.

No conclusion could be more erroneous. Recall the historical situation. The time was the early 1930s, the beginning of the Great Depression. Total demand for oil was beginning to decline after climbing steadily for more than a decade. Average wellhead oil prices per barrel in Texas had been falling steadily since 1926 from a high of $1.85 (1926 prices, remember) to $.99 in 1930, $.51 in 1931, $.83 in 1932, back to $.56 in 1933, and then the beginning of the oil price recovery in 1934 with a price of $.95. Prices for the whole United States showed the same general pattern.

The first year that the price of oil was recorded for East Texas as a separate area was 1931: $.46 per barrel. The price rose to $.94 the next year, fell to $.56 in 1933, and then rose to $1.00 in 1934.*

When the East Texas field was discovered by an independent oil company, demand was decreasing and total supply was still increasing from previous discoveries. Suddenly a very big, new, low-cost field entered the market. The inevitable market response occurred. The price of oil fell to clear the market, i.e., to restore the balance between supply and demand. The lowered price shifted production away from the older, relatively high-cost producers (whose potential output would therefore be saved for a later time) and toward the new, low-cost areas. In short, what happened would have happened even if there had never been a "major" at all.

The "facts," then, are not as the writer interpreted them. We find that even a knowledge of the historical context changes our perspective. And economic analysis leads us to see that the price drop was an outcome of the usual market forces, not the action of a cartel.

This book is in two parts. The first part introduces some ideas from economics, evaluates the oil industry's competitive behavior, and looks at its performance in supplying low-cost oil. It examines profits, investment, discovery rates, and the existence of a "crisis."

The second part looks at the growth in the government's role in the 1970s and offers an explanation for gas lines; it examines the Windfall Profits Tax and explains why the OPEC cartel has remained in fine health until recently and why "equity" among refineries costs the consumer money. The last chapter draws the policy implications of the analysis, noting both what is economically desirable and what is politically feasible.

---

*The figures in the last two paragraphs are from *Petroleum Facts and Figures*, published by the American Petroleum Institute (Washington, D.C.), 1971. The ultimate source cited by the A.P.I. is the Bureau of Mines.

PART ONE

CHAPTER ONE

Realistic Ideas
About Competition

THE OIL INDUSTRY IS REALLY composed of five sub-industries: exploration, production, crude distribution, refining, and product distribution.

The exploration sub-industry searches for and "proves" oil deposits. Exploration involves geologists, engineers, statisticians, and accountants in thousands of firms, big and small. Some firms specialize in obtaining or processing geological and geophysical data, some in drilling, others in providing equipment for drilling.

The production sub-industry extracts the crude oil from known deposits. Both small and large firms are involved in production.

The third sub-industry is crude distribution—transporting the extracted crude oil to the refineries. In the United States the chief means of transportation is pipelines.

The fourth is refining. Refineries, which number in the hundreds, are also of varying size, but until recently refining firms varied less in size than the exploring and producing firms. Refineries are set up to handle specific types of crude oil and cannot be switched easily and quickly to other types. The cost of shutting down a refinery and later starting it up again is substantial, so that all firms try to keep their refineries running continuously, even if at less than full capacity. This part of the industry includes not only the major oil companies but also many independents.

At the end of the line is product distribution, the fifth sub-industry. Thousands of firms—independents, franchised operators, leased operations—deliver the final products, such as gasoline and heating oil, to the consumer.

In each of these sub-industries there are both small firms and large firms. Some firms operate in only one sub-industry, others in two or three. A few have operations in all five sub-industries and are called "integrated firms." However, "integrated firms" means something different in the oil industry than in other industries such as steel. In other industries an integrated firm carries out all phases of the operation from raw material to final product. But in the oil industry, this simple line of vertical integration is not the universal pattern. An integrated oil firm has operations in all five sub-industries, but each stage of operation does not necessarily rely entirely on what the firm has done in the preceding stage. For instance, firm A does exploration and finds some oil. It leases the oil find to firm B, which produces crude oil that is transported in a pipeline (or ship) owned by firm C. That crude oil is refined by firm A, which sells it to firm D for distribution. Little of the total gasoline sold by any "major" (i.e., one of the top twenty oil firms) was discovered and processed through all stages by the same firm. Let us see why this different kind of integration occurs in the oil industry.

Part of the special market structure of the oil industry arises from its technological characteristics, which include these: (1) The crude oil that goes into refining is heterogeneous, but the distillates produced from the crude—heating oil, gasoline, naphtha, and the like—are homogeneous. (2) The proportions of the various distillates produced in refining can be varied somewhat, but not much. (3) Retail markets shift over time, but large modern refineries are, of course, immobile. (4) Refineries have to be designed for specific types of crude; to change the type used by a particular refinery is difficult and costly. (5) Big refineries are more efficient than small ones but must be operated near full capacity. (6) Crude is expensive to transport, except in very large quantities. Now let us see how these facts give rise to the market structure we in fact observe in the industry.

Let us start by imagining an oil industry in which each firm is fully integrated. Exxon, for example, finds its own oil, produces it, transports the crude in its own pipelines and ships, processes it in its own refineries, and sells the products only to Exxon retail outlets.

Suppose that managers of Exxon refineries are instructed to maximize firm profits. There is only one way to do this: buy crude at the lowest price and sell the output in the most favorable market. As circumstances change, some of Exxon's markets may decline or its source of crude may become more expensive. So it makes sense for each manager to be flexible. For example, a manager will buy crude from other firms whenever they are cheaper because they temporarily have more of a particular type of crude than they can use. And sometimes he will sell a surplus of his own refinery's output to other firms. If we extend this process to each stage of the oil business and each firm, the shape of the existing oil industry begins to emerge. Everybody deals with everyone else at some time or other; each seeks the best for his own firm and in consequence delivers oil at the lowest possible price.

If we further recognize the numerous opportunities to specialize within the industry, we can begin to see an industry characterized by a wide variety of firm sizes, from thousands of small exploration and geological firms to tens of thousands of independent retailers (even the retailers who use a name-brand franchise are still very much in control of their own local operations). In exploration, for instance, there are a variety of types of activity, from the very capital-intensive, high-risk frontier exploration in which the majors specialize to the relatively low-risk, low-cost, in-field exploration done by the independent wildcatters. (The wildcatter is an oil explorer who capitalizes on the experience of the bigger firms that looked in the general area before him.) Within exploration there is room, then, for a wide range of firm sizes and for various forms of specialization and levels of risk-taking.

The uncertainties in the oil industry are much greater than those in most other industries. The science of exploration is still far from exact. Geological and geophysical knowledge have not advanced to the point where one can map the subterranean surface and promptly proceed to extract oil. Mistakes are easily made. The geological gaffe of the century, for instance, was made some five decades ago by a senior geologist for a major oil company who drilled in East Texas and actually found the fabulous field there but did not recognize it.

All estimates of reserves are purely a function of looking, and looking is an expensive business, not lightly undertaken for the academic interest of knowing how much there is. The expense of exploration can vary from tens of thousands of dollars to hundreds of millions; the chance of finding oil can range from as high as one in two (but for some very small finds of dubious economic value) to a low of one in forty or fifty or more. For example, some time ago when a large Canadian oil company was beginning to explore Alberta, it sank 113 consecutive dry holes before hitting a natural gas find.

What Is a "Market"?

A market is neither a place nor an object but the interaction of economic forces, focused on two groups of agents: "suppliers" and "demanders" or consumers. Market forces are like lines of magnetic force: their effects can be felt and measured and we must act in accordance with them, but we cannot see them. The forces that get reconciled and brought into balance through market interaction exhibit their effects in the amounts supplied and demanded and above all in the price.

The origins of the forces that influence any market, whether for crude oil or shoes or newspapers, are even more numerous and diverse than the people involved in actual market transactions. Indeed, even persons who do not actively participate in the transactions within a particular period are yet in the market if they might have participated had circumstances been a little different. What goes on in the market is influenced by the circumstances and wishes not only of the active participants but also of the potential participants.

A related concept is the difference between market forces and market actors. Market forces appear to be exerted through the activity of specific actors, but to eliminate a particular actor—even a big one, like Exxon—is not to eliminate a market force but merely to change the names. The play goes on as before. Interchange a particular consumer and a particular producer and there will be no recognizable change in the market outcome. The actors in a market seldom know or care about abstract ideas of "markets"; they act in accordance with their circumstances and wishes. This is as true for markets with a few buyers and sellers as for markets with many.

Let us look at a market with many buyers and sellers. Each buyer knows *he* does not control the price of what he buys; he may assume that someone on the supply side does. And each seller knows that *he* does not control the price either; he may assume that his supplier does. But that supplier in turn feels bound by economic realities. Each market actor—whether an individual, a partnership, a firm, a union, or a conglomerate—is ignorant of his contribution to the market. He feels that all he does is respond as best he can to "market conditions"; what he fails to realize is that by doing so he helps to create the very market he observes.

Market price is determined by all those who are or could be involved in market transactions. If a consumer is angry at the high price of beef, he is as much justified in blaming his neighbors as in blaming the local meat sellers or even their suppliers. Price is determined by both demanders and suppliers. Just as suppliers compete to sell, so demanders compete to buy.

Picture a market in which there is only one seller; he has, of course, a monopoly. Doesn't that seller control the market, and aren't the consumers subject to the monopolist's power? These are not rhetorical questions, and the answers are not simple. Having a monopoly would seem to be preferable to being in a highly competitive situation. But it is a mistake to conclude from this intuitive observation either that very big firms are all monopolistic or that all monopolists make large profits. Some monopolists are small, and monopolies *can* lose money and go out of business.

Our worries about monopolies really have to do with the consumers' options. Take, as a highly contrived example, the owner of a bucket of water who faces a man dying of thirst in the desert. Such a consumer (the man dying of thirst) is said to have an "inelastic demand"; that is, since the consumer has no other options, large variations in the price asked by the seller have little effect on his demand. But if a monopolist's customers have many options, the demand for the product is said to be elastic.

A theorem in economics shows that if a monopolist's costs of production increase as his output increases (and this normally happens), it is in his interest to choose a price at which demand is elastic. And at this price, the total revenue received by the monopolist is substantially less than the maximum possible.

Let us see what this means in practical terms.

A monopolist has only two choices about pricing policy. He can fix the amount he wishes to sell but then must accept his customers' decisions on the price they will pay. Or he can fix a price but then must accept his customers' decisions on the amounts they will buy. He cannot set both; he is disciplined, to this extent, by the demand side of the market. Furthermore, the monopolist's optimal (most desirable) price is one at which any further increase would lower the demand proportionately more than the price went up. In short, an increase in price above his optimal price would lead to a reduction in total revenue, because some customers would stop buying. No monopolist can keep raising the price and keep getting more revenue.

What is objectionable about the monopolist's price is that it is higher and the quantity traded is lower than would otherwise be the case; by "conserving the product" the monopolist gains a greater return above his costs. Typically, such increased net return is shared by the senior management, who bring about the situation, and the stockholders, who finance the operation.

Consumer Options: The Example of Heating Oil

Options are the all important determinant of how pernicious a monopoly is to the consumers. Let us consider a modest but most emotional case, the use of home heating oil.

The economist distinguishes between "nominal" and "real" prices. The *nominal* price is the one we all recognize as a "price," namely, the amount in dollars the consumer pays to get the heating oil. But the nominal price is often relatively unimportant to the consumer's decision about how much to use. The actual consumer decision is made on the basis of the *real* price, or the price of heating oil *relative* to all other prices, and also the non-dollar costs to the consumer (such as inconvenience). The full price has then two extra dimensions: it must be expressed in relative terms and it must include other non-dollar costs. For example, a sixty-cent ride on the New York subway seems very cheap—until you take into account the taxes you pay to make up for subway deficits, the poor quality of the product, the discomfort and even risk of physical danger, and perhaps the eight or ten blocks you must walk to get on it. The full price of a New York subway ride is much higher than the nominal price would reveal.

If the *real* price of heating oil rises, to what extent can we adjust to escape the worst of the burden? What are the consumer's options? Everyone now knows about the usual suggestions. Lower the thermostat; use more sweaters and blankets and give up some comfort to save heating oil. Install insulating materials; exchange this cost for savings in future fuel bills. If your house is larger than you need, rent out rooms or apartments so that tenants pay some of the heating costs, or sell the house and move to a smaller place. Consider moving to a warmer part of the country, either permanently or temporarily. Or, most obviously, switch to an alternative fuel, gas or coal or coke or wood.

The point is that we do have many, many options. Most of us do not exercise them simply because, given our incomes, the current prices of the options, and our preferences, we find that buying oil, high as its current price is, is cheaper in *real* terms than the alternatives. As the real price rises more, people will adapt more, each moving in a direction he finds least burdensome.

Over time the options become even greater, since we can improve the use of alternative fuel sources, improve the energy efficiency of house heating and insulation, and even relocate a significant proportion of the population. There are many options that we currently choose not to examine seriously, but we will if the real price of heating oil continues to rise.

Markets, Firm Size, and Specialization

A neglected aspect of markets is their diversity and specialization. The best method of selling groceries in the suburbs is not the best method of selling antiques in the city; the process most suitable for oil leases will not be the same as that for timber leases or coal leases; what works best for selling medical services will not be the same as what works best for selling architectural services. Some oil refinery firms (notably the majors) concentrate on long-term contracts to ensure a smooth flow of crude at a fixed price, whereas others (notably smaller independents) specialize in the fluctuating oil spot market, a market for one-time-only transactions. The buyer in the spot market gains when the price is lower than the contract price but pays more when the price is higher. Each buyer chooses the method that he thinks will give him the best combination of price and security of supply.

The development of efficient markets for particular types of transactions is related to another matter: the appropriate degree of firm integration (or, more broadly but less accurately, "firm size"). For example, in the oil industry even the most integrated of majors does not *build* its own drilling rigs, and many do not even *operate* the drilling rigs used in their exploration. And whenever there is a serious problem with a well that cannot be controlled, all oil firms call on firms that specialize in the particularly difficult task of capping blowouts; a small market has developed in this specialized area. We see, then, that the creation of markets is intimately connected with firm decisions about what is and what is not efficient for them to do themselves.

These decisions change as three important factors change: (1) *technology*: improved technology may make it easier for firms to do certain things themselves that they previously paid others to do, thereby leading to an increase in the most efficient size of the firm, but other technological changes can lead to greater specialization and smaller firms; (2) *market size*: the bigger the market for an industry's product, the greater the scope for specialization of task and, therefore, the greater the scope for market exchanges between sub-units and the less the firm has to do everything itself; (3) *industry maturity*: the older the industry, the greater the chance that someone has discovered the gains from specialization in some part of that process, so that in the mature industry firm A can improve its productive process by handing over a particular task to firm B and buying what it needs but only when it needs it from B.

Similar factors operate in the home. The owner of an isolated farmhouse is likely to do all his household repair work himself, whereas an affluent city-dweller can easily call on a vast number of specialists instead of doing the repairs himself. But we would see a decrease in this market activity and less specialization if, for example, there were no telephones; or if technological change were to reduce the effort, time, or skill required to do a particular kind of household repair; or if the entry of new firms into the home repair business were impeded.

Adaptive Reactions to Economic Actions

Professor I. M. D. Little of England once remarked that he had been impressed in his consulting with the British government by how rigid, journalistic, and simplistic government bureaucrats were. A particular form of this has been, at least until recently, distressingly common in almost all of our own government's efforts to evaluate policy options: the failure to take into account the probable market reaction. The typical bureaucratic evaluation of a policy has proceeded on the assumption that no one reacts to a change by adjusting his pattern of behavior and by trying to devise ways of either bypassing the change or turning it to his advantage. For example, the typical evaluation of the effect of imposing a price ceiling assumes that everyone will continue to buy as much of the commodity as he did before but no more, and that the supplier will continue to supply the same amount at the lowered price.

Fortunately, some recent Department of Energy analyses have been somewhat more sophisticated than this and have tried to incorporate projected changes in demand and supply. But as an example of the shortsightedness I mentioned, consider an action by DOE's predecessor, the Federal Energy Administration. The FEA set up the crude-oil entitlements program (entitlements, which we will look at later, are rights to buy price-controlled domestic crude, which is cheaper than foreign crude) in order to make small, very inefficient refineries more competitive with large, efficient ones. Part of the unanticipated reaction was a vast increase in the number of small refineries whose sole claim to financial viability was the receipt of government largess through entitlements.

Profit: Common Misunderstandings

"Profit" is probably the most widely misunderstood term in the economic vocabulary. It is a word that generates far more emotional heat than rational light. In this discussion I will use it not according to the economist's highly specialized definition but in the usual lay sense: profit is what is left to the owners of the firm after all the costs of doing business are subtracted from income.

Either directly or indirectly it is the general consuming public that owns most oil-company stock. Those of us who own none directly are likely to own some through our insurance or pension plans. So when profits of oil firms are eroded there is probably an effect on our own retirement incomes.

Besides financing our retirement, what role do profits play in our society? The first and most important role is to signal where and to what extent society should invest its resources. If profits are higher than average in a particular industry and that condition is expected to continue, investment will flow to the industry. New firms will enter; some old firms will expand; other old firms will spawn new offshoots as talented employees see the possible gains from taking the risks of forming their own firms. If profits decline and that condition is expected to continue, then resources are withdrawn from the declining industry and are put to work where the benefit is greater.

In short, profits point out the firms where society, acting through *individual* decisions, wants its resources to be used. Members of society reveal their preferences through individual consumer decisions, which lead to profits for those firms producing best what people want and losses to those producing what people want less. Often the social critics of this process really wish to criticize other people's preferences. For example, at one time large American cars were very popular despite the disapproval of intellectuals; in Europe, too, such cars were highly prized by those who could afford to run them. One man's taste is another man's waste.

"Profits" are what is left over after all who supply the firm with materials and services have been paid. But they are not a free good to the firm; they are in fact a cost of staying in business, to say nothing of growing. The savers who provide the firm's equity expect to be repaid, and if the firm cannot do that, then they will put their money elsewhere. If *no* firm can adequately recompense the saver for saving and taking risk, he will simply spend his income. All firms, no matter how large, are implicitly, if not actively, competing for capital funds. If a firm finances a project out of its own savings, then it is forgoing the money it could earn by investing those savings elsewhere.

A common misperception is that even a competitive firm's profit is "paid for by the consumer"; according to this thinking, a big profit means that some persons are being taken advantage of, either the consumers or the employees or both. In our private dealings we may realize that *trade* is profitable to *both* parties and that the process of exchange is itself a highly productive activity, which is, of course, why we spend so much time and effort at it. When the analysis proceeds to trades between firms and individuals, this basic fact is soon forgotten. Trade is then seen as a contest between the single consumer and the mammoth corporation, between a midget and a giant.

In this context the firm should be seen as merely a device for coordinating a very large number of exchanges. Instead of a bewildering and highly complex system in which the owner of every item needed by the firm—money, labor, management, materials—bargains directly with each consumer who wants what the firm produces, the firm engages in a sequence of fairly simple exchanges. The principle still applies—both parties to an exchange gain. When one of the parties is a "firm," that firm represents *all* the persons who furnish the firm with what it needs—the suppliers of labor (employees and management), capital (stockholders and bondholders), and materials (other firms). And both parties gain.

But what if one of the parties is a monopolist; is not the other party taken advantage of, gouged?

Consider an example. When New York City suffered a transit strike, the demand for taxis jumped dramatically. Here we have a temporary "monopoly": cabs are the only public transportation in town, and municipal regulations prohibit the entry of new cabs. Suppose that the taxi drivers are allowed to take advantage of the situation by raising their prices. A person who hails a cab to take what had previously been a $4 ride from midtown Manhattan to New York University is told that the price is now $10. He tries three other cabs and is told the same thing.

His alternatives, then, are to take a cab and pay $10; walk the forty-five blocks; or cancel his appointment. If he takes the cab, he considers that his best alternative: a $10 cab ride beats walking forty-five blocks. Therefore the rider gains by the exchange, but less than he would have gained had the price been the usual $4, and (this is crucial) *only* if the probability of getting a cab and the quality of service are the same as before. The real problem that generated the increase in taxi fares, remember, was the subway strike. The increased taxi fare is merely the visible sign of the costs of the strike.

Now, what if (as actually happened) the law prohibited an increase in the price? Would the consumers have been better off? Suddenly the demand for cabs jumps dramatically. At $10 one could have gotten a cab as easily as before the strike, perhaps more easily. But at $4 there are far too many people trying to get into the usual number of cabs. Those who succeed are the strong, the fleet of foot, the rough, the callous, or the beautiful, not necessarily those who need the cabs the most. The potential riders have to endure waiting, the extreme uncertainty of success, and decreases in the quality of service (taxis that will not wait, drivers reluctant to travel to areas where they are not sure of getting an immediate return fare, many more inconsiderate drivers who decide not to work at inconvenient times, such as the rush hours).

So which is the higher *real* price: $10 for a cab quickly and with good service, or $4 for a cab after a long wait, with as little service as possible?

The profits of $10 cabbies would show enormous percentage increases for the period. But $4 cabbies might have found they were losing money by driving during this period with no fare increases but higher costs. Many would have decided not to work at all. Would the consumer really have been better off if cabbies were not making profits but there were fewer cabs to ride?

Let us view the matter of "profit" another way. Imagine two

firms in the same industry. All you know is that one is profitable, the other is not. If you were seeking a firm in which to invest your savings, you would of course choose the profitable firm over the unprofitable one. And if you were seeking employment, you would undoubtedly prefer to work in the profitable firm.

But suppose you were going to buy a product that both firms make. You compare the two versions of the product and find them very similar in price and in apparent quality (though you cannot be sure about the *actual* quality). All other things being equal, you would probably conclude that you would stand a better chance of being satisfied by the product of the profitable firm. You might suspect that either the more profitable firm's product is better and people have discovered that fact or the firm can produce the same quality as its competitor for less. Either way, the profitable firm is more likely to survive the test of time. Consequently, in the future you would be more likely to be able to get new parts for the product, to get it repaired, and even to resell it.

If we conclude, then, that profitable firms are preferred to non-profitable firms by investors, employees, suppliers, and even consumers, we have cause to wonder at the vehemence of the opponents of profit and to question their arguments.

There are two sources of negative reaction to profits. First, if a firm is a monopolist, it can earn larger profits than other firms as long as it can prevent the entry of new firms and new competing products. The same is true if a group of firms get together to form a cartel, that is, agree to act as a single firm.

The other source of negative reaction to profits is size, pure and simple. If a firm is large, that is, is engaged in an extremely large number of exchanges, then any profit, no matter how small relative to gross income, will appear enormous. The larger oil companies quote their profits in hundreds of millions of dollars. But the profit numbers are big because the amount of business these companies transact is enormous. We see here a statistical illusion. The large numbers make the uninformed think that all the firm's suppliers, employees, and consumers are being taken advantage of, and that if profits were smaller, suppliers would get more for their labor and materials and consumers would pay less for the end product.

Consider actual oil-company profit rates for gasoline. For thirty years the profit rate has been less than 5 per cent of the selling price; for extended periods it has been less than 3 per cent. In late 1979 the oil company's profit was between 3 and 4.5 cents per gallon. Since the "average" U.S. driver was traveling 12,000 miles per year in a twelve-miles-per-gallon car, what was going out of the driver's pocket into oil-company profits was $30 to $50 per year. The elimination of all gasoline profits would gain the individual consumer very little.

Competition, Monopoly, and Cartels

What is wrong with a monopoly? Most of the usual antipathy toward it derives from the notion that a monopolist will receive a higher price and a bigger profit by restricting his output, i.e., by *conserving* the product, in this case oil and gas. But the monopolist is not a free agent; he too is disciplined by the market, for there must be a consumer demand for his product. The monopolist can raise his price, but he must then accept consumer reactions to that price. There are strict limits to how high a price a seller can charge before he brings a *decrease* in total revenues and in profit because consumers refuse to buy.

But how effective is the consumer market's disciplining of the monopolist? As we saw earlier, the greater the consumer's options, the more the monopolist is penalized with reduced total revenues by any price increase he might impose. But there is a strong tendency for customers to ignore or fail to recognize their options. This is partly because the weighing of alternatives is time-consuming and so not to be engaged in lightly.

A while ago we looked at cab-riding during a New York City strike of subway and bus workers. During that strike, a transit service that carries more than 3.5 million passengers per day was suddenly halted. New York City is infamous at the best of times for having a nearly perpetual traffic jam and for having almost no parking facilities. Theoretically, it was a situation with no realistic options; the city would have to close down until the strike ended.

But from the very first day of the strike, people exercised a wide variety of options. Car pooling, private buses, and multi-passenger rides in taxis were an obvious first step. Parks, courtyards, alleys, lawns, and empty lots were used as parking spaces. Riders discovered that a New Jersey system provided a subway link from downtown to midtown Manhattan. Some commuters from outer Long Island and Connecticut slept in their offices and in hotels. People came to work by bicycle or on foot—walking, running, or roller-skating. A flotilla of boats was organized to ferry people from uptown, Brooklyn, and the Bronx to downtown Manhattan. One enterprising soul built a rickshaw and sold rides. Some used helicopters.

The result was that on the first day of the strike, between 61 and 88 per cent of the normal work force in various industries arrived in the city ready for work. By the third day attendance was near normal for that season of the year. But the adjustments were not over. Further plans were developed in case the strike went on for some weeks. They included "clearing houses" for commuter rides; more buses; greater use of the rivers and the development of river taxis; staggered working hours; a four-day work week with the off-day staggered through the week; more use of bicycles, motor scooters, and motor bikes; arrangements for showering in the city for those who walked or ran in.

This example of how New Yorkers coped with the seemingly impossible illustrates a basic point: If the price is high enough, people will quickly and even cheerfully consider a wide variety of options that they otherwise would not contemplate. Furthermore, they might find they actually prefer the new way to the old. When the transit workers returned to work they discovered that the demand for their services had decreased, even though the price charged was only slightly higher than before.

A monopolist must consider the long-run adaptability of both consumers and other producers who react to his high prices. The potential loss of revenues over the long haul is a powerful inhibitor of monopolistic behavior.

We live in a dynamic world of change and reaction. Monopoly profits without strong legal or political support do not last. And the monopolist of a given product is always in competition with other products, new entrants into the industry, and, most important of all, consumer choice.

Cartels, even in the absence of anti-trust laws, are even more vulnerable than monopolies and are in greater need of political or legal support to survive. For in a cartel, the gains each member can realize by increasing his production while the others restrain their production are enormous. As soon as one or two members cease to restrict their output and begin to capitalize on the gains from "cheating on the cartel," the cartel is doomed.

Government Support of Monopolies

But an important factor is missing from this discussion. Our arguments seem to say that we needn't worry much about monopolies and cartels, for they are restrained by market forces and are doomed to failure, often quite quickly. The unmentioned factor, one that changes the prognosis, is government. While monopolies that rise out of competition (like Kodak's monopoly of domestic camera production) are of no great concern, legally supported monopolies are. It may at first sight seem surprising that the U.S. government would support monopolies, but that will be a surprise only to those whose experience is limited to our more recent history. Traditionally, all governments have favored monopolies; the "granting of monopoly rights" was until this century a prime source of governmental revenue. If we extend our notion of the legal protection of monopolies to the granting of only partial or limited monopolies, then governments are universally involved. The procedure is seen in import quotas, tariff protection, exclusive franchises, "consumer protection" laws designed to limit entry, and the full-scale legal monopolies found in public utilities.

Why should governmentally backed monopolies be a source of concern if the monopolies and cartels that arise out of competitive behavior are not? The reason goes back to the idea of the market, which recognizes the force of the potential entry of new competitors and the potential loss of demand by consumers. With a legally protected market, the firm faces less pressure to innovate, increase efficiency, and respond to consumer preference. Legal monopolies do not have to worry about potential entrants, and that one apparently simple fact makes a vast difference in behavior. Indeed, some monopolies or cartels manage to acquire legal backing to force the consumer to use their product; for example, laws in some states prohibit the attachment of non-Bell Telephone equipment to Bell telephone lines.

If one reads carefully any history of an industry, it soon becomes clear that to the extent that monopolistic elements are sustained, the fine hand of governmental support is observed. An excellent example of this is the oil industry itself. A close reading of John Blair's *The Control of Oil* shows time and time again the governmental actions that encouraged and aided monopolistic behavior, and John Blair was certainly no friend of the oil industry. However, it would be wrong to categorize the industry as a simple monopoly. Despite some governmentally encouraged monopolistic behavior, it is a highly competitive industry.

The problem here is that the government frequently has the power, and often exercises that power, to obtain for a competitive industry the sort of benefits that firms could otherwise attain only by becoming an effective cartel; in short, government does for them what they cannot do for themselves. If an industry wants prices raised and the impact of the market's discipline reduced, then it must prevent the entry of new firms into the industry. With municipal governments, licensing is the favorite ploy. This is how New York City's taxi industry limits entry and obtains higher incomes for cab drivers. Cab driving in the *absence of government regulation* is as competitive an industry as one is likely to see. Yet cartel pricing can be obtained and enforced by the law, which ensures the survival of the cartel policy, though without such a law there is no earthly hope for a cab cartel in the city. A partial measure of the extra income gained through restriction of entry is the fact that a license now costs about $68,000 on the market. New York riders can also testify that licensing does not insure good service, honesty, or even an adequate knowledge of the city among drivers.

Why government should create monopoly pricing in a competitive industry is another question. The answer is partly that it is a political trade in which the economic costs are not considered. A standard political trade involves the exchange of a group's political support—in money, votes, or campaign workers—for a licensing restriction, an import tariff or quota, or subsidies. While the restrictions are usually billed as being in the public interest, the inevitable effect is to raise the price and lower the quality to the consumer—and to increase the probability that a particular candidate will be elected to public office.

Recall that the oil industry comprises a vast, heterogeneous array of firms that vary greatly in size, specialization, and degree of integration. Each firm's self-interest is very much its own. However, a firm will often find that it is in its interest to join some coalition of firms seeking government support for a tariff or a quota, a subsidy, aids in kind, and so on. The coalitions joined vary from issue to issue and are seldom industry-wide. Legislation that benefits one group will often penalize another. For example, oil import quotas used in the 1950s benefited domestic—especially *small* domestic—producers but penalized refiners relying on imported oil. Therefore, when we speak of legislation we miss the essential nature of the problem if we do not recognize the heterogeneity of oil-firm interests.

During the 1950s, some firms were seeking cheap imported oil rather than the high-priced domestic oil and some of the majors had an enormous self-interest in supplying that cheap oil. But a very large number of small domestic-production firms scattered all over the United States had an interest in limiting imports and thereby protecting their own market. *Import quotas* benefited these firms at the expense of those that could supply cheap imported oil. Majors were on both sides of the fence.

The *oil depletion allowance* enables a firm to reduce its taxable income by a certain amount to compensate for using up its stock of oil. The allowance was substantially reduced from its original terms for the majors but retained in full for smaller production firms. The *entitlements* program forces domestic producers to subsidize importers of the more expensive foreign oil; it is a massive income transfer from some majors to other majors and independents. Contrary to popular opinion, the *prorationing of production* across firms, which was developed in the 1930s, was designed to keep in production an incredible array of relatively inefficient small-volume producers. This was done at the initial expense of the majors and the larger-scale independents but ultimately at the expense of the consumer.

Almost all the legislation applied to the oil industry benefits some at the expense of others; even in the unusual case in which all firms benefit, some benefit more than others. Again contrary to popular opinion, the one systematic bias in all legislation is to favor a large number of small but vote-important firms—whether in exploration, production, or retailing—at the expense of a small number of large firms. Two consistent facts emerge: the disparity of firm self-interest and the ephemeral nature of political coalitions in the oil industry.

The oil industry is not unique in receiving government support through such means as the depletion allowance. Farmers, for instance, have received the benefits of price supports, subsidies, tariff barriers, and import quotas as much as anyone; by these means, achieved through various political exchanges, they have managed to keep their incomes at a much higher level than would otherwise would be possible. They have achieved the effects of limited entry and cartel restraint of production by persuading the government to buy nonproduction, i.e., pay some farmers not to produce.

Medical doctors and dentists, teamsters and independent truckers are among the groups that have stood gratefully under the government's protective umbrella. Licensing and restricted entry have been the main tools of achieving incomes higher than unrestrained competition would support. In short, government legislation has done for each group what it could not do for itself: protected it from competitive pressure from new entrants and migrating consumers and thereby enabled it to achieve higher average prices and incomes than would otherwise be possible.

Nevertheless, many of these groups would be fully competitive in the absence of the government umbrella. Even under the umbrella there is often fierce competition. This is true of the oil industry: it is competitive even with governmental involvement but could be much more competitive. The basic effect of such legislation is to produce only the illusion of stability in prices and supplies, while in fact increasing supply rigidities, and to prevent the market from weeding out the small, inefficient producers.

OPEC: A Genuine Cartel

The various quasi-academic discussions of the extent and degree of collusion among the international oil companies before 1971 pale into pedantic quibbling before the startling clarity of the effect of a genuine cartel. I will, of course, say more later about OPEC—the Organization of Petroleum Exporting Countries—and its effect on the domestic industry. The point to be made in this discussion of the competitiveness in the industry is simply that OPEC is a genuine and highly effective cartel.

Earlier I claimed that cartels without legal support were short-lived. But OPEC has so far shown few signs of collapsing. Its apparent long-term stability derives from two sources: it is as much a *political* group as an economic one, and the U.S. government has done a great deal to support and maintain its power. This of course was not Washington's intention, but it was the effect.

The U.S. State Department, with the connivance of the British, Belgian, and German governments, helped to make the cartel effective in the first place. The stated U.S. rationale for such incredible economic behavior in 1970-71 was to seek stability in the Near East; if this really was the intention, it reflects an amazing level of naïveté. The more astute reason for the Europeans was to protect high-cost European domestic coal production. For example, Britain, which is heavily dependent on its exports for survival, put a tariff on imported oil before the OPEC era so that the domestic oil price would be equivalent to the corresponding price of domestic coal; in short, the exercise was one of support for coal miners, a very powerful coalition in the United Kingdom.

Since 1971 the United States has ensured, albeit indirectly, that oil *imports* were subsidized to about $3 per barrel. It has done this by setting up the oil entitlements program, restraining domestic production, increasing regulatory constraints, encouraging consumption, and imposing mandatory consumption rates. No wonder that the United States' reliance on Arab oil has increased from about 1.9 per cent of domestic consumption at the beginning of the decade to nearly 22 per cent at its end—an elevenfold increase during the great Arab-instigated price increases. That demand for the highest-price oil should rise dramatically relative to demand for the lower-price (domestic) oil is an amazing development that could occur only as government pursues political aims at almost any economic cost. No wonder the OPEC cartel has not only survived but prospered.

CHAPTER TWO

The Structure of the Oil Industry

A MAJOR OBJECTIVE OF the first chapter was to provide a framework for examining the actual structure and performance of the oil industry, as we will do in this chapter and the next. The facts about the industry must be interpreted within a general understanding of how the oil market functions.

Unquestionably, some oil firms are big; their sales and net worth are expressed in billions of dollars. But big is not necessarily bad. The U.S. government, General Motors, and the Teamsters are big; the United States is a very big country. To get some perspective on the numbers, we must not only compare the oil industry with other industries but also examine trends over as long a period as possible.

Economists have traditionally judged an industry's degree of competitiveness on two grounds: structure and performance. The analysis of *structure* tries to determine what conditions are necessary for a small group to be able to wield enough economic power to increase significantly the returns to the industry. For example, if there are only two firms in an industry, then it is much less difficult for the firms to act in some collusive manner than if there are twenty firms. Other matters that are potentially important for a structural analysis include the number and degree of joint ventures and the nature and extent of vertical integration and horizontal integration.

In the analysis of *performance*, it is assumed that if there is a significant degree of monopoly in an industry, then the effects of that monopoly should be measurable in the performance of the industry and the returns it earns. However, this task is difficult, for energetic and innovative non-monopolistic firms can be highly profitable and monopolistic firms can be losers. Other aspects of performance are the rate of innovation and the extent to which the innovation improves efficiency; but it is difficult to devise the right yardstick for measuring these aspects.

In examining the oil industry's structure and performance, particularly over the last eight years, we must remember that the industry is now much more tightly controlled by government than ever before. The Department of Energy, the Environmental Protection Agency, the Occupational Safety and Health Administration (OSHA), Congress, and other federal bodies have much more to say about who gets oil or gas and how much it costs than any other group with the exception of OPEC. Much government legislation has the effect of raising costs, hindering the entry of new firms into the industry, and reducing the flexibility of prices. Sometimes the effect of government action is one that otherwise could be achieved only by the formation of a cartel that could keep out new competition, restrict output, and raise prices.

In this chapter and the next, however, we will examine the competitiveness of the industry as though there were no governmental involvement. The question will be: Is the industry competitive when compared to other industries?

The Level of Concentration

Our first task is to consider the extent to which a small number of firms might be able to dominate the industry by controlling the market, that is, supplying almost the whole of the demand for that industry's products. Although such firms are still subject to market forces, especially through consumer preferences, they are not immediately subject to direct competition from other suppliers. Firms in a competitive industry are penalized for mistakes very quickly, but monopolistic firms are penalized mainly in the long run.

The economist's traditional and rather simple measures of the *potential* for control are the so-called *concentration ratios*. These ratios show the proportion of total industry sales accounted for by the largest four, eight, or twenty firms. The twenty-firm ratio is somewhat academic, since there is very little scope for any effective collusive action among that many firms. Effective collusion requires extensive agreements on market shares and pricing, and these have to be monitored constantly. The temptation for one or more firms to cheat on the agreement increases rapidly with the number of firms involved.

Since the concentration ratios shown in Appendixes 1 and 2 for various parts of the oil industry are not very informative if there is nothing with which to compare them, Appendixes 3, 4, and 5 have been added to show figures for other industries. However, since the tables show ratios for separate *sub-industries* of the oil industry and the ratios for the *whole* of other industries, the comparison will overstate the relative concentration in the oil industry.

(A note about the appendixes: The tables and graphs included there provide some evidence for the statements in the text. However, because most readers—and I include myself—do not enjoy wading through a sea of charts, I have tried to make the text complete in itself, without the appendix materials. I will often make reference to those materials, and the figures will be there for anyone who wishes to study them.)

The figures for concentration ratios in crude production (Appendixes 1 and 2) provide our first illustration that a superficial examination of data can be very misleading. We see that the ratios at the four-, eight-, and twenty-firm levels all increased substantially from 1955 to 1970. Does this mean that the industry is becoming ever more collusive?

The 1970 figures, which show the heaviest concentration ratios, are 31 per cent for the top four, 50 per cent for the top eight, and 68 per cent for the top twenty. These seem high—but are they? If we look at the figures for other industries (Appendixes 3-5), we find almost no concentration ratios any lower. Only for plastics in 1967 (Appendix 3) are the ratios lower than those for oil production. The levels for such diverse products as electric lamps, chewing gum, typewriters, vacuum cleaners, greeting cards, and liquor are far higher. Furthermore, the concentration ratios at several levels of the productive process in oil and gas do not vary greatly throughout the domestic phases of the industry. In short, the oil industry (1) is at most only modestly concentrated at any level of its operations and (2) is one of the least concentrated industries.

But these oil figures go only to 1970. Does the fact that they rose steadily between 1955 and 1970 point to the monopolization of the industry and a continuing increase in the concentration ratios?

We need to look at reasons for the increase in concentration during the 1960s. Two factors were at work. First, this was the period of the greatest shift from on-shore to off-shore drilling. In the early stages of a costly new exploratory effort, it will be the largest firms that finance and lead the high-risk, high-cost development. Later, as the technology spreads to the industry as a whole and as, therefore, the costs fall, more and more smaller operators will join in.

The second factor is more important. From the mid-1950s to the beginning of the 1970s, the costs for domestic production and exploration were far greater than the corresponding costs for imports. Consequently, during this period domestic exploration diminished; and so, therefore, did the number of domestic exploration firms. Also, as the industry shifted toward a greater reliance on imports, concentration inevitably rose, since a few major companies dominated that portion of the supply. The Middle East crises arose before the anticipated flow of smaller firms into this sector of the market could begin to move concentration ratios back toward their earlier and lower levels. Therefore the rises in concentration levels after 1955 were the natural outcome of economic events, in particular the diminishing relative importance of domestic production.

Another question is raised by the concentration figures: Have the rankings of firms within the concentration ratios changed? Has there been entry into and exit from the list of the top twenty and the top eight?

The lists for 1955 and 1970 (Appendixes 6 and 7) show that six of the companies listed in the top twenty in 1955 were not there in 1970—and, of course, there were six newcomers on the 1970 list. (Some of this movement in and out occurred through mergers.)

Within the top four, Gulf displaced Standard Oil of California. Within the top eight, Getty was a new entrant (even to the top twenty), and Atlantic Richfield, seventh in 1970, had been six-teenth in 1955. Continental Oil, which had been eighth in 1955, was only thirteenth in 1970. In short, the relative market shares of the companies are changing. If we were to look at the shares within regional markets, this change would be even more pronounced. Since change in market shares is what cartels strive hardest to prevent, this evidence is significant to the question of collusive tendencies.

Something else that will put the "monopoly potential" of the structure of the industry into perspective is an analysis of violations of Section 2 of the Sherman Antitrust Act. This section concentrates on the structure of the market that a firm is accused of monopolizing or of forming a conspiracy to monopolize. Appendix 8 summarizes successful prosecutions under Section 2. It shows that the successful antitrust cases were those in which the conspiring firms had more than 75 per cent of the market, where there were fewer than four large competitors, and where for substantial periods no new competitors entered the market. These characteristics do not apply to the oil industry.

Charges of Collusion

Some critics of the oil industry cry collusion whenever they encounter an aspect of the industry that does not fit into a highly simplistic competitive theory. But careful analysis often reverses the conclusions of intuition and casual observation.

For example, in a 1974 book entitled *Highway Robbery: An Analysis of the Gasoline Crisis*, Fred Allvine and James Patterson claimed that the majors were collusively squeezing the independent refiners out of the market. Their charge was enhanced by a 1974 study by the Independent Gasoline Marketers' Council (IGMC) showing that, for those of their members surveyed, market share fell 17.1 per cent between 1972 and 1974.

However, the membership of the IGMC consists of fifteen of the larger independents that have 16 per cent of the total number of independent outlets but only 2 per cent of total sales. Many of the members eschewed long-term contracts for oil and instead bought on the spot market—a risky decision but one they make at their own initiative.

The various spot markets in oil and refined products are important parts of the continuous daily interaction between supply and demand. The spot market for oil is rather like the local all-night convenience store: it's not where you do your major shopping, but it's very useful for occasional purchases. Price in the spot market fluctuates in response to changes in the supply-to-demand ratio. When supply exceeds demand, the price drops. A company that chooses to buy the bulk of its oil in the spot market rather than by long-term contract takes a risk: when the price is lower than the contract price it gains, but when the price is higher it loses.

The independents that chose to buy through long-term contracts had no trouble maintaining their market shares; those that took their risks with the spot market lost out when supply fell relative to demand and the spot price rose above the long-term price.

Further, of the top fifteen gainers in retail market shares by state during 1970-73, four of the top five, five of the top ten, and six of the top fifteen are non-branded independent oil companies. The top gain among these fifteen was 20.1 per cent in Alaska by an independent; the smallest gain was 2 per cent, earned by three majors.

This evidence shows that the Allvine and Patterson charge of conspiracy by the majors to force out the independents is unfounded. We see here a good example of how the *selection* of data can lend plausibility to a misguided charge.

Another example of an ill-founded claim of conspiracy is the charge by the Federal Trade Commission that the majors used the (now eliminated) oil depletion allowance to "squeeze independent refiners." Under the depletion allowance an oil company could reduce its taxable income earned from crude production to compensate for the depletion of its inventory of oil in the ground. The FTC staff suggested that since through use of the depletion allowance the effective tax rate on crude oil production profits was reduced, therefore the majors would try to shift profits over from the refining part of their operation to crude production so as to benefit more from the lower effective tax rate. To do this they would raise the price of crude.

This hypothesis may seem plausible, but let us test it against the facts. First, if the majors are to set the price of crude, they must dominate its supply and restrict the entry of other firms into exploration and production. We have just seen that the former requirement does not hold. As for the latter, most critics of the oil industry recognize that the entry of firms has not been measurably impeded. The most recent figures available from the Census Bureau of Mineral Industries (1977) show 16,800 exploration firms, 3,587 oil production service firms, 2,180 drilling contractors, and about 6,000 firms involved in crude extraction.

A second reason why this FTC charge seems ill conceived is that according to the estimate of Professor Richard B. Mancke as reported in *Vertical Integration in the Oil Industry* (edited by Edward W. Mitchell), sixteen of the top seventeen integrated majors would have found the process speculated on by the FTC unprofitable, even on the basis of the FTC's own data. Most of the major firms are not self-sufficient in crude and must therefore purchase crude for their refineries from others, i.e., from independent producers. Professor Mancke's conclusions from the same data differ from the FTC's because he actually examined the hypothesis in detail instead of relying on its plausibility. Mancke estimated that in order for it to work, the majors would have to be more than 93 per cent self-sufficient in crude and would have to dominate the supply of crude, which they do not. Only Getty was self-sufficient; the highest self-sufficiency figure for the rest of the majors was 88 per cent, and most were well below that figure. (The data are presented in Appendix 9.)

The foregoing is another example of the tendency for some critics of the industry to subscribe to a simplistic and highly abstract theory of "competition" and then, when some industry behavior or structure is not explained by that naïve theory, to attribute it to collusion. The preferred approach would be to test a theory of collusion directly, examining the actual evidence for or against it.

Are Joint Ventures Collusive?

A striking example of this confusion of thought is found in the attitude toward joint ventures. For many industry critics, the very existence of joint ventures is evidence of collusion.

Among the fundamental factors in economic decision-making are the problems of risk, uncertainty, and change. Joint ventures are an attempt to minimize risk. They are not unique to the oil business. They are likely whenever there is some probability that a contemplated costly project will be profitable but also some probability that it will result in a loss. Examples include everything from the underwriting of new bond issues to the production of movies and the funding of America's Cup boats. Joint ventures reduce both the risk and the potential return to each participant in an uncertain undertaking.

How can joint ventures as a way of reducing risk be distinguished from joint ventures as a form of collusion? First of all, we might note that, especially in this country, if firms are colluding, then there is little profit to be gained from engaging in joint ventures also. In fact, it would be risky to draw attention to the colluding firms; one might reasonably expect a collusive group of firms never to engage in joint ventures with one another.

Before we look at joint ventures in pipelines, consider an analogy. A small firm makes some prepared foods that it sells to restaurants in a distant town. The foods must be transported in a refrigerated truck, but the volume of business and frequency of delivery are such that the firm would never fill a truck on any trip. Other firms in the town face a similar situation. Would they not engage in a joint truck-buying venture? This is precisely how pipelines were financed and built.

Pipelines have been under the regulatory supervision of the Interstate Commerce Commission (ICC) as common carriers. This means that a pipeline operator must provide service at nondiscriminatory rates if capacity is available and that the service is subject to continuous ICC surveillance. Consequently the probability of collusion in the pipeline part of the industry is small.

Is Cooperation in Refining Collusive?

Another aspect of the oil industry that, though really evidence of competitiveness and efficiency, is often misunderstood as evidence of collusion, is cooperation in refining. At two key stages in the processing of oil a smooth flow of homogeneous product is required: when crude oil enters a refinery and when the distilled output is distributed. Refineries work best when operated continuously, at a steady rate, and at near capacity. Given the uncertainties and vagaries both of the oil supplies of the type required by a particular refinery and of the demand for that refinery's output, an efficient and flexible method of exchange of oil and oil products has developed—and is crucial.

Oil and its products are redistributed very quickly and efficiently through three interacting mechanisms: spot markets, exchange agreements, and processing agreements. These mechanisms are not now subject to regulatory controls. Spot markets we have already discussed. Exchange and processing agreements are cogently explained in *Competition in the Oil Industry* (by William A. Johnson and others) as follows:

> Exchange agreements are also, in effect, a highly efficient form of transportation. Rather than ship crude oil and refined products long distances, both integrated and independent companies have found it mutually beneficial to swap oil among themselves. A skilled crude and product trader is one of the most valued employees of an oil company. Using only a telephone, he is often able to "move" crude from Texas to California or product from Seattle to Bangor in a few hours. And if he is exceptionally skilled he is able to do this at minimal cost to his company. Exchange agreements are widely used for both crude oil and refined products. Because of these agreements, a significant share of the crude oil refined by an integrated company may actually come from other producers or owners, while a significant share of the product sold by the company may come from other refiners.
>
> Processing agreements are less common. Under these agreements, oil companies pay others to refine their oil. Most processing agreements involve newly built refining capacity. A refiner's capacity is often expanded in discrete jumps because of economies of large scale operations. To make full use of capacity

until marketing operations also expand, refiners will agree to process crude oil owned by others for a specified period of time. While exchange agreements are a means of transporting oil as little as possible, processing agreements are a means of utilizing refining capacity as efficiently as possible. Both types of agreements benefit consumers by permitting lower prices. . . .

In preparing this chapter, we have talked with officials of a number of oil companies. All major oil companies with whom we have spoken have made crude trades with independent refiners and producers within the past year. They have also made product trades with independent as well as major oil company refiners. The attitude among the major oil company officials was a willingness to trade with anybody as long as the price was right and the oil was of the proper type and quality. Similarly, all small refiners queried have made trades with major oil companies, and those who wished to trade but did not possess any crude oil of their own have usually been able to purchase the crude necessary to make exchanges. None expressed substantial displeasure with the treatment now received from their major oil company trading partners [William A. Johnson *et al., Competition in the Oil Industry* (Washington, D.C.: George Washington University, 1976), pp. 78, 79, 81, 82].

A *competitive* market generates these agreements as the most efficient method of adapting to rapidly changing circumstances.

Suppose, as we did before, that we are able to regroup the industry into independent, fully integrated firms; each firm explores for oil, extracts the crude, transports it to the refinery, refines it, and distributes the products only through its own outlets. Now imagine that each manager at each stage of production in each firm is told to maximize profits. Given the industry's technological, economic, and geological facts of life, the same structure and market arrangements we currently see would soon develop. Some firms would find their retail outlets growing slowly or shrinking with changes in population and domestic trade; others would find their markets growing rapidly. Some firms would have to buy crude oil from other firms in order to keep their refineries busy; others would have a surplus to sell. Some would find crude in excess of their needs or of a type they could not use; others would not find enough. Some would discover they needed new transportation facilities, others that they had an excessive capacity. Before long a complex structure would develop, with firms of various sizes and degrees of integration engaging in specialization, spot markets, and exchange and processing agreements.

Is Horizontal Integration Collusive?

The last major structural issue is horizontal merger or horizontal integration. Some years ago an article in *Fortune* magazine assessed the chief characteristics that distinguished the successful companies from the unsuccessful ones. Only one major characteristic was found: Successful firms had a broad conception of their product and industry. The successful firm was not just in sports cars, but in transportation; not just in running shoes, but in sports equipment and gear; not just in computers, but in data management; not just in oil, but in energy. The successful firm is one that can adapt quickly to changing circumstances, one that is always seeking new products, new ways of using old products, better ways to meet consumer preferences. The more broadly defined the company's interests, the more likely it is to recognize how research efforts in one area of its work can produce benefits in another area.

From the firm's short-run point of view, a broadly based definition of its line of products provides a hedge against errors in forecasting supply and demand. The greater the risk inherent in dealing in any one product, the more important it is for the firm to diversify. Consequently we should expect the oil companies to acquire access to alternative fuel sources and to try to develop alternative energy supplies. This they have been doing for years.

Predictably, they have been criticized for this response to risk. They are charged with attempting to increase their monopoly power by buying coal firms. This would follow only if (1) they had a monopoly to begin with and (2) the coal market was such that buying into it could increase the oil companies' monopolistic control. Appendix 10 lists the top twenty firms in coal reserves (1970) and Appendix 11 lists the top twenty in production (1974). These tables show, first, that the coal industry does not have high concentration ratios, and, second, that the oil-company penetration is not sizable. For example, while Exxon's coal company is listed as fifth in reserves, it was not among the top twenty in production for 1974. The highest-ranked producing coal company owned by an oil company is Continental's Consolidation Coal, which ranked second in production in 1974. But Continental, though a large company, certainly does not rank in the Exxon/Shell/Texaco league. Although concentration ratios in coal are still relatively low, they have been increasing since 1955. The reason is that until recently coal was a declining industry; an increase in concentration ratios was to be expected as the weaker companies left the field. The data are shown in Appendix 12.

We see, then, that the major oil companies do not have a sufficiently dominant role to be regarded as potentially collusive in oil, and the probability of collusion is even less in coal. My conclusion is that the purchase of coal companies is the outcome of portfolio diversification in the oil industry, not of collusion or the attempt to increase monopoly power.

Motives for Charging Collusion

While these arguments taken one by one seem plausible and consistent with the general tenets of economic theory, still we hear frequent authoritative complaints that the industry is "collusive" in some sense. Does not all this smoke mean there must be a fire somewhere?

A point to keep in mind is the self-interest of all parties involved. Recall that the reputations—not to mention the financial rewards—of many people would be enhanced greatly by the successful prosecution of an oil-company conspiracy. The career of an economist at the Department of Energy or a lawyer for the Justice Department or Federal Trade Commission or a politician from a non-oil-producing state would benefit greatly from a successful case against Exxon or Shell or Mobil for significant violations of DOE regulations or the Sherman Antitrust Act. And so any apparently unusual action by the oil companies is regarded with suspicion.

I am not making a moral judgment, nor do I wish to "defend" the oil companies, which after all are operating in *their* self-interest.

The companies are often guilty of transgressing laws and should be punished when they do so. I wish merely to make the point that self-interest and the search for gain and glory are not restricted to businessmen and university professors.

CHAPTER THREE

The Performance of the Oil Industry

THE PRECEDING CHAPTERS HAVE NOT answered the consumer's immediate concerns about whether oil will continue to be available and at what price. These questions will be the subject of this chapter. We will look first at the prices of oil and its products, turn next to what has been happening to the supply of oil, and then move on to the emotionally charged subject of dealer margins and company profits.

Gas and Fuel Oil Prices

As we saw in chapter one, the "real" and the "nominal" prices of gasoline and fuel oil are quite different. Nominal prices are what the consumer pays at the pump or to the fuel-oil dealer; they have been rising for some time and have indeed risen a lot. Real prices are the nominal prices adjusted for inflation, that is, for the effect of the overall general increase in all prices, wages, and incomes. Such prices are called "constant dollar prices" and must be related to some index base. Some point in time is chosen at which the nominal price is considered to be equal to the real price; all other nominal prices are then rescaled to allow for changes in the value of the dollar from the base. This enables economists to see how the price of a particular item changes relative to the change in overall prices. If the *real* price does not rise or even falls, then, other things being the same, we will not expect actual consumption of that commodity to fall, even if the *nominal* price rises dramatically.

41

Domestic real prices for both gasoline and heating oil fell steadily from 1956 to 1972. There were some years of increase, but the overall trend was down. Between 1973 and 1974 the real price increased 21.7 per cent for regular gasoline and 42.7 per cent for heating oil. But after that the real price of regular gasoline fell again until 1980, when it started to move up once more. If we compare an index of gasoline prices to an index of food prices for 1950 to 1980, we discover that the cumulative increase in gasoline prices was exceeded in every year but two by the cumulative increase in food prices (see Appendix 13). Since it is also true that real incomes rose during that period, albeit slowly over the last six years, we should not be surprised to find little reduction in the consumption of gasoline.

This is not the whole picture, however, since the long-term adjustments are made on the basis of anticipated *future* prices, not existing prices. It is now reasonably clear, as it was not in 1973 and 1974, that future real prices for gasoline and oil can be expected to rise. This anticipation held down somewhat the rise in consumption that would otherwise have occurred as the real price of gasoline fell and real incomes rose between 1975 and 1978. Of course, if real prices of gasoline and heating oil *rise* and real incomes *fall*, then there will be a further decrease in actual consumption.

Currently, U.S. energy use per dollar of GNP is starting to decline, and the rate of decline appears to be increasing. The decline can be expected to accelerate over the next few years, especially if the real price of energy rises. Time is needed to adjust to higher prices, and we are only beginning to realize the consumption savings brought about by conservation efforts initiated during the past few years. The decline in use increases as more and more energy-saving efforts become effective. (See Appendix 14.)

An explanation of gasoline usage since 1967 must reflect (1) the effect of the Environmental Protection Agency measures that led to decreased fuel efficiency, (2) the effect of the real price rise in 1973-74, (3) the offsetting effect of the rise in real incomes, and (4) the effect of government-sponsored fuel-conservation measures. Miles per gallon fell steadily from 1967 (the base year) to 1973 and then began an increase as steady as the previous decline. More strikingly, miles traveled per car rose until 1973, fell sharply in 1974, and continued to fall slowly after that. (The figures are in Appendix 15.)

These trends and the path of real prices for more than a generation put the current prices into perspective. Current *real* gasoline prices are still below the highs attained in the late 1950s and are much lower than in the 1930s. If the population could tolerate the real gasoline price that prevailed in 1957, surely the same real price in 1979 offset by much higher average *real* incomes and with greater public benefits to the poor and the unemployed cannot be said to impose an unconscionable burden on anyone. The current situation is one in which we are back to prices we paid before, but with much higher incomes with which to pay those prices; in short, we are still much better off than we were in the 1950s.

The real price of heating oil is another matter. Real prices in 1978 were about 40 per cent higher than 1956 real prices. But, as is being evidenced more and more, this increase in real prices has generated a boom in the insulation and fuel-saving business. Furthermore, since real incomes have increased on the average by 56 per cent over approximately the same period,* and since, again, public support to the lowest income groups in our society has increased vastly over the past twenty-five years, the actual increase in real prices ceases to seem an unprecedented and insupportable burden and becomes merely an item of concern and a stimulus to conservation. With a 56 per cent increase in real income and a 40 per cent increase in real heating costs, fewer hours of work are needed to heat one's house in 1979 than were needed in 1956. And a 40 per cent increase in the real *price* of fuel oil does not mean a 40 per cent increase in heating *bills;* that increase will be much less as people add insulation, substitute other fuels, or simply turn the thermostat down.

Another aspect of the situation comes to light if we compare U.S. and German price increases (exclusive of taxes) for gasoline

---

*Table No. 714, *Statistical Abstract of the United States, 1978* (U.S. Department of Commerce, Bureau of Census).

and fuel oil. Germany's increase exclusive of taxes was 125 per cent from October 1973 to June 1979. Because Germany had a low rate of inflation and the Deutsche mark was one of the strongest currencies, this price rise is a clearer reflection of the actions of OPEC than the U.S. price increase, which was 146 per cent even after strenuous efforts to hold it down. A big contributor to the U.S. nominal price rise was domestic inflation.

Europeans pay much more than Americans for gasoline not because of OPEC but because of their governments' stiff taxes on gasoline consumption. For example, while there was no U.S. increase in direct taxes on gasoline between 1973 and 1979, the rise in Italy was 160 per cent. France, the United Kingdom, and West Germany had increases between these two extremes. Federal tax levels in 1979 were 55 per cent for Germany, 42 per cent for the United Kingdom, 72 per cent for Italy, 66 per cent for France, and 15 per cent for the United States. (The figures are in Appendix 16.)

Oil Search and Oil Reserves

Exploration and lease development are investment expenditures, and investment expenditures are undertaken only when one *expects* to earn a profit. If it were likely that the price of oil would rise but also that any future profit from oil would be heavily taxed, then a low rate of investment would be understandable. Similarly, if OPEC prices rose but domestic prices were not allowed to rise accordingly, and this situation seemed likely to continue, then once again a low rate of investment would be likely.

Popular myth holds that except for some off-shore leases there is no oil left to be discovered in the United States and so there is no use talking about higher prices and profit incentives. This myth is shared by many who write books on the energy problem: they studiously ignore oil exploration in favor of conservation and the search for exotic fuels and exotic processes—exotic in the sense that existing technology does not enable us to use these alternative fuels and processes at a cost comparable to the cost of using fossil fuels.

There *is* tremendous scope for further exploration in this country. However, the costs of exploring will be higher than they have been in the past and are likely to continue to rise with the currently available technology. The easily found fields in Texas, Louisiana, Oklahoma, and other highly accessible areas have been found. We now must reach for smaller pools and deeper pools in less accessible areas. However, the very perception of these cost rises will further stimulate the search not only for feasible alternatives but also for technological change to lower the cost of oil search and extraction.

Another puzzling point for the consumer is the conflict in statements issued about long-term reserves. A government agency's estimates may even conflict with governmental policy. For instance, in an article in 1979* Congressman Clarence J. Brown noted that D.V.E. McKelvey, a former director of the United States Geological Survey, estimated while still in office (July 1977) that there are likely to be more than 100 billion barrels of recoverable oil not yet discovered, in addition to current known reserves. (Cumulative fuel use from 1859 to 1976 was 122 billion barrels; the current annual production rate is roughly three billion barrels.) The corresponding figures for natural gas were 484 trillion cubic feet to be discovered (in addition to reserves of approximately 200 trillion cubic feet) and a cumulative consumption of 495 trillion cubic feet. In addition, there are preliminary indications of truly enormous quantities of natural gas in geopressurized zones in the Gulf Coast region.

A study by the Energy Research and Development Administration (ERDA, an agency that became part of the Department of Energy in 1977), "Market Oriented Program Planning Study," came to conclusions similar to McKelvey's. (The director of that study and McKelvey, then director of the USGS, were both dismissed. The director of the ERDA study was quoted as saying under oath during a Senate inquiry that he was removed as director because his results conflicted with the "conventional wisdom.")

Let us consider the exploration possibilities on the basis of

---

*"Energy and the Economy: The Moral Equivalent of Surrender," *Commonsense*, Fall 1979.

publicly available information. The scope of exploration is most easily seen for the off-shore areas. The portion that has been leased is as yet inadequately developed, and it amounts to only 2 per cent of the available area. In 1975, over one-fifth of the natural gas and one-sixth of the domestic crude produced came from the Outer Continental Shelf, even though the U.S. government's lease sale in the summer of 1977 for the very promising eastern Gulf of Mexico was the first in four years. Alaska is being explored and developed rather slowly, but the scope for future exploration is vast, perhaps matched in extent only by Texas at the turn of the century. Even Texas, Louisiana, Oklahoma, Kansas, California, and Michigan still provide rich opportunities for exploration.

Gasoline and fuel oil, even at current OPEC prices, are still the best energy buys in most uses; yet, ironically, exploration is ignored or dismissed by nearly everyone—except the industry itself. In 1977 there were 16,000 exploration operators in business, the vast majority of whose efforts were concentrated on-shore and not in Alaska.

Another myth is that once a reservoir is discovered and its maximum efficient rate of recovery is determined, then no more decisions about output are left except to close it down when the oil (or gas) is gone. That is far from the truth.

First of all, the optimum rate of initial recovery is not just a technical concept but an economic one: higher prices justify higher production rates. Secondly, as recovery proceeds over time, the costs of recovery rise as the pressure of the oil drive drops and as equipment wears out or becomes corroded. Another reason for rising costs as ever greater amounts are extracted is the possibility of using a variety of recovery-enhancement techniques. The stage at which a well is shut down because it can no longer produce oil at market prices leaves a lot of oil in the ground; the average in the past has been greater than 60 per cent. As the real price of crude oil rises, old shut-down wells once more become economically viable; they should, and in an unimpeded market would, be reactivated. In short, at sufficiently high real prices (which we are currently approaching), an enormous amount of known oil can be extracted from old reservoirs, often using old wells. The number of such reservoirs and wells is legion. And so there are substantial—albeit costly—opportunities for increasing supply even without any new discoveries.

As a historical example of what effect the removal of price controls has on supply, consider the removal of wartime price controls in 1946. Wellhead prices for oil and gas rose 84 per cent within three years and then stabilized, but the number of producing wells in the United States doubled within a decade. Other factors besides the price increase played a part, but the point is that, notwithstanding the other factors, little increase in the number of wells would have occurred had the crude oil price been held to the wartime level.

President Nixon's price freeze in 1971 began a decade of controls, either direct or indirect, on crude oil prices. And, after the dramatic increases in OPEC prices in 1974, the controls developed into an ever more complex system of multiple crude oil prices. The analysis of prices is further complicated by the fact that the prices represent differences in the quality and availability of different types of crude.

OPEC and Prices

The price increases initiated by OPEC show clearly what an effective cartel can do. While OPEC as an organizational structure was formed in 1961, ten years were required along with the (possibly inadvertent) aid of the U.S. Department of State and other foreign governments to generate an economically effective cartel. This status was finally achieved at a meeting in 1971 in Teheran.

The first few years of the cartel were spent in moving net oil revenues from sales of the final products back to the suppliers of the original low-cost crude—the OPEC nations. After that, the only way to increase revenues was to begin supply restrictions. This OPEC did by maintaining output levels in the face of increasing world demand and by raising the price. The OPEC nations succeeded beyond their wildest dreams in that the United States in particular, through its own government's new role in oil market control, managed to *increase* reliance on OPEC sources.

The distinction between real and nominal prices is strikingly evident in OPEC crude oil prices. The popular assumption is that OPEC prices took a big jump in late 1973 and have risen continuously ever since, sometimes in big jumps and sometimes in small. This is a basically correct perception of the *nominal* prices (shown in Appendix 17).

The path of *real* OPEC crude oil prices is somewhat different (the graph is in Appendix 18). Late 1973 brought a huge increase in real prices, about fourfold, but thereafter the picture is somewhat different: real prices *fell steadily* until mid-1979, when OPEC reachieved part of the initial real gain. In short, a good portion of the post-1973 price increases in OPEC oil is merely the reflection of our own high domestic inflation rate and the consequent rapid rate of depreciation of the dollar against foreign currencies.

The Path of Domestic Oil Prices

The high points for domestic real crude oil prices at the wellhead were the years 1949 ($3.23 in 1967 dollars) and 1957 ($3.31 in 1967 dollars—recall the Suez crisis), after which real (and even nominal) prices fell steadily until 1973 ($2.88 in 1967 dollars). (The figures are in Appendix 19.)

After 1974, average domestic prices became a creature of the government through the price-control policies administered by the Federal Energy Office and later by the Department of Energy. The government defined a series of "categories of oil sources," each with a different price. This main categories were Lower Tier, Upper Tier, Alaskan, and Stripper (which was unregulated). The average price of unregulated "Stripper Oil"—that is, oil from wells producing less than ten barrels a day—was approximately equal to the average landed price of OPEC oil, or $6.57 per barrel (1967 dollars) in 1979. The prices of "Lower Tier Oil" (also called "Old Oil") were meant to hold constant the returns on production begun by 1975; in 1979 the Lower Tier price was $2.57 (1967 dollars). Oil produced in excess of 1975 base levels and in new fields was priced at the "Upper Tier" level; in 1979 that price was $5.69 (1967 dollars). This higher level was meant to act as a stimulus to new production and discovery while still holding domestic prices below the world price. (The figures are in Appendix 20.)

Let's reconsider these prices within a more complete historical context. In 1974, the Lower Tier price was set at approximately the 1973 real price, and it later was reduced by 17 per cent. The reduction in real terms from the 1969-70 price of $3.30 was 28 per cent (that includes an adjustment for the reduction of the depletion allowance for the majors).* However, the 1973 real price did not itself represent a market-clearing price, since crude oil had already been subject to price controls of varying stringency since Nixon's price freeze, begun in 1971. If we wish to maintain production from existing wells over time, then real prices must rise, not fall.

As for the higher Upper Tier prices ostensibly set to encourage new production, the figures show that after 1975 real prices fell. The situation is worse because even in real terms the cost of drilling a well and costs per foot of oil exploration have been rising substantially. Indeed, between 1974 and 1977 the real average cost per well-foot drilled increased by 32 per cent (Appendix 21). So while real returns on new exploration are falling, the real costs of exploring are increasing, a situation that certainly does not encourage risky new exploration.

These comments help to explain the disappointing percentage of overall production for Upper Tier oil, the strength of the unregulated Stripper percentage, and the decline in production of Lower Tier oil. (The figures are shown in Appendix 22.)

Let us now try to get some historical perspective on exploration and production, demand, and imports. We should keep in mind the broad shift in supply conditions since 1946. In the first few years after World War II, the domestic crude industry was built up rapidly. The incredibly vast and productive reserves of the Middle East started to come into production during the early 1950s, so that by the mid-1950s domestic production was being replaced more and more by relatively cheap imported oil. These pressures on domestic production, especially on the numerous small-scale

---

*Edward W. Erickson et al., "The Political Economy of Crude Oil Price Controls," Natural Resources Journal, October 1978, p. 795.

domestic producers, led to attempts at import controls during the late 1950s. The controls had only limited success; the tendency to provide exceptions for East Coast refineries and for the chemical and plastics industries proved irresistible. It seemed clear at that time to those in the industry that the relatively high cost of most domestic oil sources compared to Middle East sources would continue indefinitely.

But after the 1971 Teheran conference, when the OPEC cartel came into being, it was apparent that OPEC would push more vigorously to extract all the rent due on its low-cost oil. So even in 1970-71 it was clear that the price differential between U.S. and OPEC oil would be eliminated, if not reversed. And after 1973, the relative prices of domestic and OPEC oil were indeed suddenly and dramatically reversed.

The Time Path of Exploration

This brief history of prices and price expectations is mirrored in the time path of exploration. On-shore sources were developed intensively in the United States while these sources were relatively cheap, until 1956; then development slackened off dramatically while the Middle East countries were the low-cost producers, till the 1970s. Development in the United States became attractive again in 1971 and 1972, as it became clear that OPEC was intent on raising its prices, but was delayed and impeded by the Nixon price freeze. (The chart of footage drilled is in Appendix 23.)

The proportion of exploratory wells to development wells drilled (Appendix 24) reflects the same economic conditions. The proportion fell during the 1950s, rose in the late 1960s, and has fallen steadily since 1970. The post-1970 decline reflects the fact that the net real returns on exploratory drilling, especially since the depletion allowance was reduced for the major firms, have been falling. The costs of exploration are rising: exploratory depths are increasing (Appendix 25) and the amount of oil discovered per foot drilled is falling (Appendix 26). A *higher* real return is needed to make exploration attractive, but instead the real return is lower. Oil companies therefore prefer to develop familiar areas rather than take the risk of exploring new fields. This is not to say that no new fields have been explored and developed, but with benefits not nearly so great as outsiders think, much less has been done than would have been done within an unimpeded market.

A good indicator of the industry's reactions to anticipated future prices is the number of rotary rigs in operation (these are used in exploration) and the number of seismic-crew months worked each year. (The charts are in Appendixes 27 and 28.) This evidence, too, shows that U.S. exploration declined greatly between the mid-1950s and the early 1970s, as the industry moved away from relatively expensive U.S. sources to relatively cheap foreign sources.

The Time Path of Reserves

Our discussion now enables us to evaluate the path over time of the ratio of oil reserves to oil production. The U.S. reserves-to-production ratio (shown in Appendix 29), or reserves expressed in terms of "years of supply," stayed roughly constant at a level of between ten and twelve years until the early to mid-1960s. At that time a slight decline began. This reflected the accumulated impact of the falloff in exploration that had started a decade earlier. When conditions stimulate the oil companies to engage in new exploration again, it will take nearly a decade—perhaps even more—for the reserves-to-production ratio to start to recover.

The reserves-to-production ratio for natural gas is quite different and has been predicted, charted, and debated for decades. The gas reserves-to-production ratio is a steady downward path (Appendix 29). This is the outcome of Federal Energy Regulatory Commission efforts to keep the price of gas below market levels: demand is stimulated, but supply increases are impeded. The situation would have been much worse, and the decline in the reserves-to-production ratio far more dramatic, if gas discoveries were not also a by-product of oil exploration. This is lesson number one in how to create a shortage: hold the price below the market level.

The time path of crude reserves the reader can by now anticipate: an early buildup of reserves as an outcome of exploratory effort in the post-war period, followed by a decline in reserves as an outcome of the decrease in exploratory efforts beginning in the mid-1950s. Because of the time delays involved in exploring and proving reserves, we should expect the reserve figures to change later than the change in exploration. In fact, reserves peaked in the mid-1960s. (The chart is in Appendix 30.) Considering the very slow rate of leasing of the off-shore areas and Alaska and the meager exploration incentives on-shore, this decline in proven reserves is no mystery. Nor does it suggest that reserves are about to be used up with little chance of being replaced. The decline is now seen to be the natural result of recognized economic and political forces. The geological reality is not that there is no oil and natural gas left to discover but that the cost of finding them increases steadily.

A recent myth about exploration is that oil discovery rates are dramatically down. The success rate for exploratory drilling has in fact improved moderately during the past decade. There are two major reasons for this improvement: better technology, and a recent shift in exploration away from the riskiest areas toward those that are relatively safe but low in expected yield. This change to lower-risk exploration is one of the outcomes of price controls, which lower the anticipated returns from exploration.

Consider next how the rate of reserve additions compares to the rate of production (Appendix 31). Additions to reserves fell ever more rapidly after the mid-1950s, but production continued to rise slowly; this was yet another natural outcome of the shift from domestic to foreign sources because of the changed relative prices. One other aspect of the industry is very apparent from the figures on additions to reserves: the enormous uncertainties involved. Additions within a year or two fluctuated by as much as 1.5 billion barrels, even though exploratory efforts varied relatively little.

The time paths of production of oil and natural gas (Appendix 32) demonstrate two important facts. First, while off-shore production has become relatively more important since the mid-1950s, it is still a small part of the total. Second, production peaked in 1971, the year of the Nixon price freeze and the beginning of price controls on domestic production.

Because net imports are the balancing factor between domestic supply and demand, the final part of the story is easily told. (The graph showing import percentages is in Appendix 33.) With the relatively low price of foreign crude beginning in the 1950s, the percentage of imports should have been expected to rise; this is indeed what happened. By the late 1960s, total U.S. imports had reached approximately 20 per cent of domestic demand. Before the 1970s, the Arab proportion was less than 10 per cent of imports or about 2 per cent of domestic demand. But after 1972 the Arab proportion rose rapidly. In 1975, the first full year after the first major OPEC price increase, the Arab proportion rose to 22.8 per cent of imports, which now supplied approximately one-third of the domestic demand; this means that U.S. reliance on the main instigators of the price increase had nearly quadrupled. By 1979 the reliance on the Arab sources had risen to nearly 22 per cent of *total demand,* and nearly half of the total demand was met by imports.

We see, then, the following interesting situation. A group of nations forms a cartel to raise prices. Instead of having to restrict output in order to achieve the increase, the cartel merely restrains the rate at which it increases its output. In response, one of the major consuming nations promptly increases both its absolute and its relative demands from the cartel, and thereafter it increases that dependence annually. Under these conditions no cartel could fail. For when the demand for a cartel's product increases continually and substantially, the normal internal strains generated when cartel members have to share a smaller level of output are eliminated.

A further ironic twist: the chief source of relatively cheap oil outside the cartel, which could have been instrumental in putting competitive pressure on the cartel, was our own domestic industry. It is incredible that the United States, faced with the sudden emergence of the OPEC cartel, did not switch its demand to the cheaper sources but instead increased its demand for the higher-priced cartel oil. This seemingly bizarre behavior can be explained only in political terms, for it certainly makes no economic sense. That these actions can, in fact, be explained in political terms offers little economic solace.

Dealer Margins, Real and Nominal

A charge recently leveled at the oil industry is that gasoline "dealer margins" have increased dramatically over the past few years. The dealer margin is the difference between the price at which the dealer buys the gasoline and the price at which he sells it. This difference is meant to cover all the dealer's costs. If the dealer's costs rise, the dealer margin has to rise also. These margins were held down first by the Nixon price freeze, then by the Cost of Living Council controls, and finally by the controls of the Council on Wage and Price Stability and the Economic Regulatory Administration of the Department of Energy.

Estimates of dealer margins by the National Congress of Petroleum Retailers for the period July 1975 to June 1979 are generally in accord with the Department of Energy figures for that period, though the Department of Energy estimates are somewhat lower. The figures from the National Congress cover two additional years, however. Using this more complete set of figures, one obtains an average margin in nominal cents per gallon of 10.8 cents in July 1975, rising slowly to 12.4 cents by June 1979. The real value of the margin fell over this period by 4 per cent. During the period January 1975 to January 1979, total volume in gallons increased by only 13 per cent. Part of the increase in volume, moreover, was due to the growth of gas sales by discount stores; this growth offset a decline in the volume of business handled by independent dealers. For example, a survey of independent dealers in Georgia showed that 79 per cent had volume declines that averaged 26 per cent over the four years 1973 to 1977. Another indication that gasoline dealers are certainly not in a growth industry is that since 1972 more gas stations have closed than have opened. In 1976, for instance, there were 5,357 more closings than openings; 5,399 more closings in 1977; and 4,691 more in 1978.

For dealers in residential heating oil, Department of Energy figures show that the *nominal* margin increased from 6.5 cents per gallon in 1975 to 12.2 cents per gallon in 1978. The corresponding real increase was 57 per cent over the four-year period, or about 14 per cent a year. The dealer margin as an average percentage of total selling price rose from 17.2 per cent in 1975 to 24.7 per cent in 1978. Data limitations make it unclear at this time to what extent this gain represents a catching up from prior price-control restraints. Furthermore, while the real gain in the dealer margin is substantial, it is also not clear to what extent the industry faced a real increase in operating costs. What is clear, however, is that home heating-oil dealers have done much better in real margins than their gasoline-distributing colleagues.

The Path of Profits

We turn now to profits. The profit rates we will examine are for a large sample of producing oil companies, a somewhat smaller sample of refining companies, an average for manufacturing (not including oil), and an average of the top oil-producing companies. The profit rate is defined as the percentage return on net worth (after taxes) on an annual basis. (The data are shown in Appendix 34.)

The first item of interest in these profit comparisons is that on the average the oil companies have been similar in profitability to non-oil industries. Producing companies have done better, on the average, than refining companies.

Secondly, the average profit rate has fluctuated between 11.8 per cent and 35.9 per cent for producing companies and between 9.9 per cent and 22.1 per cent for refining companies. The year-to-year changes can be dramatic; for example, refining profits increased by an average of 48 per cent between 1946 and 1947 and decreased by an average of 40 per cent between 1948 and 1949. The fluctuations in profit rates of individual firms, even the top twenty-five, are even more dramatic (Appendix 35). For one short period when prices rose rapidly because of OPEC actions, 1974-75, percentage gains in net income varied from a low of 2.4 to a high of 140.8 per cent, but over the same period percentage *decreases* in net income varied from a low of 3.6 per cent to a high of 60.1 per cent. Clearly, what benefits one firm's profit may harm another's. The average change during this period was a 28.4 per cent decrease. Over the two years 1973 to 1975, the average decrease in net income was 3 per cent, but the low and high percentage gains were 34 per cent and 163 per cent, and the low and high percentage decreases were 14 per cent and 76 per cent.

So far, the numbers seem to show high profits in the oil industry. Gains of 34 to 163 per cent sound grossly excessive to the gasoline purchaser. Even though percentage declines also appear large, from 14 per cent to 76 per cent, the numbers are smaller than the numbers for gains. We might conclude at first glance that, while there is a lot of variation over time and across companies, the oil companies appear to be doing very well indeed. But these statistical facts presented without interpretation are likely to be misleading.

There is a simple trap in these percentage-change figures. Suppose you bought a share of stock for $100, and the first year your dividend was $1, the second $2, the third $1, and the fourth $12. Your profit for the second year showed a 100 per cent increase— but then for the third year it went down 50 per cent. Here is the first important lesson; a 100 per cent increase followed by a 50 per cent decrease brings the profit level back exactly to where it started. Similarly, the 76 per cent decrease in oil-company profits mentioned above would offset a prior increase of 317 per cent. In the fourth year, your profits rose at the "obscene" rate of 1,100 per cent; clearly you are a pernicious parasite on society and should have these grossly excessive profits heavily taxed.

But let's see how you, the shareholder, would view the matter. First, over the four-year period your return totaled $16 on $100, an average return of $4 per year, which is just 4 per cent. Suppose your personal income is such that in the low-income years your marginal tax rate is 15 per cent, at a dividend rate of $4 per year it is 20 per cent, and at a $12 dividend rate it is 50 per cent. Now consider which dividend income pattern you would prefer, the fluctuating $1-2-1-12 or $4 per year. The four-year marginal tax adds up to $6.60 for the fluctuating dividends but to $3.20 for the steady 4 per cent. True, income averaging can be used to reduce the tax on the fluctuating dividends, but the tax would still be higher than the tax on the steady 4 per cent dividends.

The point is that the mere statement of even very large percentage gains in profits and dividends does not necessarily imply that the recipients are on the *average* earning extraordinary returns. Profits by their nature fluctuate widely. A bad year followed by a good year, or even only a modest year, will produce a wild percentage increase, while a good year followed by a bad one will yield a dramatic percentage decline.

Another way of looking at profits is to consider the amount a producing company earns on a barrel of oil after production costs are subtracted. (Remember that the profit is in a sense a cost: it repays the shareholders for the use of their money.) The average figures given in Table 3-1 show the total earnings per barrel of oil in the industry during the heady years of huge OPEC price increases.

Table 3-1
AVERAGE OIL PRODUCER EARNINGS
(Dollars per Barrel)

	1974	1975	1976	1977	1978	1979*
Income from Products	12.23	13.78	14.84	16.42	17.36	22.96
Production Costs	12.16	13.60	14.25	15.72	16.64	21.98
Net Return	0.07	0.18	0.59	0.70	0.89	0.98
Net Return as % of Average Price	.5%	1.3%	4.0%	4.3%	5.1%	4.3%

*Average of first three quarters only.

SOURCE: Data from *Monthly Petroleum Review* (New York: Merrill Lynch, Pierce, Fenner & Smith), November 1979, p. 38.

If we divide these per-barrel figures by forty-two (the number of gallons in a barrel) we have the per-gallon figures for gasoline. Even during 1979, the producing firm's net return per gallon was $.023. The quarterly net returns for 1979 were $1.05 per barrel in the first quarter, $0.98 in the second, and $0.90 in the third.

With a delivered price of fuel oil at nearly a dollar a gallon (including taxes) and of gasoline at well over a dollar a gallon (including taxes), the high 1979 profit figure of $.023 per gallon, even if expropriated entirely and given to the consumer, would not be an exciting sum. Most consumers would probably not even notice the price reduction created by profit expropriation.

Although average oil-company profits rose between 1974 and 1976, that period was one of difficult adjustment for producers.

They were caught between the economic effects of unilateral OPEC price rises and the U.S. government's domestic efforts to hold prices down. Cash flow was eroded for many companies, and the apparently high profit figures were due in part to the statistical mirage of the reevaluation of inventory stocks.

Consider next the oil companies' capital expenditures for future production—that is, for exploration, lease acquisition, and the like. From 1971 to 1979 these expenditures increased nearly fivefold, from $4 billion to $23 billion. Though these figures are in current dollars, they represent a remarkable increase in real expenditures also. More important to note is that exploration held steady in nominal dollar terms from 1955 to 1971 (and therefore fell substantially in real terms) and then began to rise rapidly, first for lease acquisitions in 1973 and then for physical exploration in 1976.

The lease figures show a muted response relative to the general increase. For example, U.S lease acquisitions rose from a 1970-71 annual average of $700 million per year to $3.6 billion in 1973 but promptly declined even in nominal terms over the next four years, to recover only in 1979.

Worldwide exploration and production have increased eightfold since 1963, mostly after 1970. Even after adjustment for both inflation and the real increase in exploration costs, the actual increase of activity in the last few years can reasonably be called an explosion. The U.S. share of total worldwide exploration and production expenditures has fallen from nearly 80 per cent in 1950 to 50 per cent in 1980. And, in addition to increasing less than the worldwide average, U.S. exploration has changed to favor relatively low-risk but low pay-off efforts, in particular by the large independents and the majors.

Another factor in the complex relation among profit levels, risk, and the rate of exploration is that exploration is a very risky and a very specialized business. Because of this, lenders are not eager to finance it. Exploring firms must rely on internally generated rather than borrowed funds much more than they would have to do in a less uncertain and less specialized situation. When they *are* able to borrow, the lenders expect the firms to bear a much higher proportion of the total risk than usual. Therefore the level of current cash flow is a vital element in determining the rate of exploration in the short run.

In Appendix 36, Mobil's figures are typical for the whole industry: capital and exploration expenditures tend to follow profitability levels because of the close correlation between profit flow and cash flow. Furthermore, while many firms use their own resources to a considerable extent in financing exploration and production, many have to borrow part of what they need. Success in such borrowing and in raising new equity funds depends in large part on a firm's track record; if that is weakened, the firm's ability to borrow and therefore its exploration efforts are seriously inhibited. In short, a good profit record facilitates borrowing for risky but possibly highly profitable ventures.

However, cash flow is the really key item in the flexible low-cost financing of risky exploration ventures. Efforts to hold down domestic prices have as one immediate effect the depressing of cash flow, and this impedes the ability of firms to engage in risky investments, which is exactly what oil exploration is.

The last item in our examination of profits is the charge that the oil companies are investing not in oil and energy development but in just about anything else. Mobil, for instance, bought Montgomery Ward and the Container Corporation.

Such purchases are usually included in a catchall category of expenditures called "Other" in statistical summaries. In general, the "Other" category amounts to less than 4 per cent of total capital expenditures. Given the types of investment that are in "Other," the percentages are remarkably low. Furthermore, the category includes a wide variety of investments in energy and energy-related fields, everything from coal mines to solar research firms. Purchases like Mobil's acquisition of Montgomery Ward therefore make up only a small percentage of an already small percentage.

A well-run company, one that expects not only to survive but to prosper, will diversify its portfolio to include investments outside its industry. It does this (1) to minimize fluctuations in cash flow, especially when unexpected events impede the firm's cash flow from its major endeavors, and (2) to provide a hedge against fluctuations in profit. This is not an extraordinary procedure. Even individuals who are engaged in occupations where the income flow is highly variable develop alternative lines of employment in order to have a safety net when demand is slack.

While we now see that the oil-industry investment in non-oil enterprises is small and in non-energy-related fields is, on the average, minuscule, what if the charge that oil companies were investing in everything but energy had been correct?

If the oil companies are not investing in oil, our immediate concern should be to ask why. Why do they think oil is going to be unprofitable? Why are they getting out of the industry? The popular myth has it backwards: it assumes that oil companies will buy up non-oil firms when their oil profits start to rise! This is irrational. If oil profits are potentially greater than non-oil profits, the company invests in oil.

Finally, let us look at profits of the five largest majors for the first three quarters of 1979 as reported in the *New York Times*. With the background we now have we can put the figures into perspective. At least we know that the figures for any one year should be seen in the context of a general understanding of the industry and an awareness of its history.

The profit increase was reported in this way in the *Times* (November 26, 1979): "So far this year, the five largest American oil companies have earned a record $7.9 billion in profits, or 75 per cent more than they made in the first three quarters of 1978." While these statements are factually correct, they mislead by offering no context for the figures. How did profits compare to profits in other industries? What rate of return on investment do the profits show?

Table 3-2 shows that this "75 per cent" increase in the profits of five companies was part of a general increase that brought third-quarter oil-company profits up from 2.8 per cent *below* the average for manufacturing industries in 1978 to 4.5 per cent above that average in 1979. A less volatile figure from year to year is the percentage return on total assets. In this the oil companies improved their position from 1.8 per cent below the average for manufacturing industries to 1.7 per cent above it. An examination of Appendix 34 will show that these figures are merely a single round in the continual cycle of position changes in the rate-of-return stakes between oil and other industries. (The figures in Table 3-2 and those in Appendix 34 cannot be compared directly because the two use different definitions of profit rate.)

Table 3-2

AVERAGE PROFIT (NET RETURN) OF
OIL AND MANUFACTURING INDUSTRIES

	THIRD QUARTER, 1978	THIRD QUARTER, 1979
Return on Investment		
Oil Industry	13.3%	21.3%
Manufacturing Industries	16.1	16.8
Return on Total Assets		
Oil Industry	6.2	9.9
Manufacturing Industries	8.0	8.2

SOURCE: Data from American Petroleum Institute, Washington, D.C., as cited in the *New York Times*, November 26, 1979.

The five target oil companies are very big enterprises, and any total figure cited about them will be big, whether it be barrels of oil sold, number of employees, exploration expenditures, or total profits. We have seen, too, that the percentage gain in a situation of recovering from lower levels of profit looks much bigger than the corresponding decline that brings profits back to where they had been; if the profit for year B increases 100 per cent over year A, a 50 per cent decline in year C would bring the profit back to the year A level. We can now more correctly conclude from the figures cited that the big oil companies during 1979 managed to recover from a relatively low profit position in 1978.

When we recognize further that the OPEC pricing action should have created the most favorable profit situation for domestic firms that they have seen in thirty years, we should wonder not why 1979 profit figures are high but why they are so low.

SUMMARY OF PART I

Our discussion has covered a lot of territory, some of it in detail, the rest only in the broadest terms. Let us pull together a few of the major lines of thought, beginning with a summary of the history of the domestic oil industry.

After World War II ended in 1945 and price controls were removed, U.S. domestic oil exploration and production entered a decade of tremendous expansion. There was an escalating demand for fuel to replace coal in homes and industries, for motor fuels, and for oil to feed the nascent petrochemical industry, especially in plastics.

Real prices of gasoline peaked in 1951 and then began a reasonably steady decline that lasted for twenty-one years, till 1972. Prices declined because supply increased faster than demand and because for the first part of this period the real costs of production and distribution fell.

During the early 1950s it became clear that the best opportunities for exploration were abroad, especially in the Middle East. The Middle East's unexpectedly massive discoveries soon affected the profitability of the domestic industry as enormous reserves were built up and oil was discovered and produced at very low cost. Domestic exploration decreased at an accelerating rate.

By the mid-1950s the influx of cheap foreign oil had greatly eroded domestic producers' share of the market. The many small producers were able to get legislation inhibiting the imports of those majors involved in foreign production. The effectiveness of the import-quota policy was hindered, however, by the many politically motivated exceptions made to supply the petrochemical industry and to meet the Northeast's demand for fuel oil. The oil supply had become politicized.

As domestic exploration dropped off, output increases slowed. Finally, in the late 1960s, domestic ratios of production to reserves peaked and began the inevitable decline after a decade's decrease in exploration.

During this period and even more during the 1970s, the real cost of exploration rose. In this sense, the era of "cheap" U.S. domestic oil was over. The real cost rose first of all because the easier, cheaper, and bigger on-shore sites had already been found, so that new oil had to be discovered off-shore, or in deeper, more inaccessible, and more difficult terrain, or in smaller pools.

Then came OPEC, and the relative prices of domestic and foreign oil were reversed; once again domestic production became relatively cheap. But the government stepped in again, this time in favor of the short-run interests of consumers rather than the interests of small-scale producers as before. The domestic industry was held back from responding to its favorable price situation by the government's short-run policies of holding down consumer prices through controls and of inhibiting profit gains by domestic producers. Both actions were politically popular but involved considerable economic costs, not the least being the inhibition of a full-scale expansion of domestic exploration and production, though the expansion was still considerable.

We now see that the reduction in domestic output began as an economic response to the cheap and abundant oil discovered in the Mideast but became more and more a function of political and bureaucratic control. The experienced "shortages" were political facts, not geological facts.

One of the points to be seen in this brief history is that oil firms are neither wholly hostile nor wholly favorable toward government intervention; they decry government action that restricts their options but praise action that inhibits the entry of others, lessens competitive pressure, and makes economic life more secure—even if less profitable. The latter they see as necessary steps to "maintain a viable competitive economy." But that is a shortsighted, wrong view. A "viable competitive economy" is achieved by vigorous enforcement of the laws of contract and the laws against fraud and by governmental restraint in responding to the political temptation to provide subsidies, direct and indirect, to consumers, workers, and even owners of capital.

Another lesson is that the oil industry is by no means a monolith with a single set of agreed objectives. It embodies many conflicting claims of self-interest: importers versus domestic producers, large-scale refineries versus small, inefficient ones, small-scale on-shore wildcatters versus large-scale off-shore oil gamblers, independents who buy on the spot market when supply is restricted versus independents who choose long-term contracts, West Coast producers versus Southwest producers, and so on.

Crucial questions of policy import are whether the domestic industry as it has developed since World War II has been competitive and whether it has been extraordinarily profitable. The short answer to the question of profitability is no and to competitiveness is a decisive yes—if we have a more sophisticated understanding of markets than that contained in the typical undergraduate's first course in economics. Vital factors of the oil market that are not included in simplistic versions of competition are the role of risk and uncertainty in decision-making; the specialization by firms in different types of exploration; the recognition that exploration firms are specialists in a particular type of risk-taking in which their own net worth and cash flow help to determine the level of exploration they can undertake; the recognition of change and the time period needed to adapt to change; the ambivalent position of firms on governmental intervention; and the willingness of government to provide, through legislation, benefits of cartelization that a group of producers on their own could not achieve.

In our examination of the structure and performance of the industry we learned that the industry is competitive, that the majors do not dominate the independents, and that profit levels expressed as a percentage of capital invested or net worth, while volatile, are on the average no larger than those in manufacturing in general. Further, when expressed as a percentage of the price of a gallon of gasoline, for example, profit rates are less than 5 per cent and often less than 3 per cent, so that the potential consumer gains even from expropriation of all profits are insignificant. Moreover, profit is not just an excess available for the taking; it is the necessary return to the equity-holders for their willingness to forgo consumption by saving and to take the risk of getting little or no return if the firm does not earn sufficient income.

The 1971 success of OPEC's efforts introduced to the world market a cartel large enough to set international prices. The creation of the OPEC cartel and the substantial price increases it dictated gave the United States a major policy opportunity. The political U.S. response was a serious economic mistake.

The best economic response would have been to let the market respond. The market response would have provided the biggest increase in non-OPEC supply, the fastest reduction in quantity demanded, and the greatest dispersion of trades of crude oil, thereby increasing the difficulties for OPEC in policing the actions of its members. The more demand is centralized into a relatively small number of national markets where governments implicitly guarantee delivery to their own national consumers, the greater the economic and political leverage given to OPEC.

Let us suppose that the United States had made the plausible market response to the OPEC action. After OPEC unilaterally raised the price of its oil in 1974, the average price of domestically produced oil would have risen quickly, but not immediately, to the world price (long-term contracts would have slowed the rise). As I estimated in late 1973 and early 1974, the domestic price for gasoline (regular) to clear the market at that time—that is, to balance supply and demand in the face of OPEC's restriction in supply—would have been in the region of fifty cents to sixty cents.

The accounting profits of domestic—now low-cost—producers would have risen immediately and dramatically, and, more important, the expectation of future real profits would have risen dramatically also.

Refiners would have shifted from OPEC oil toward domestic oil and oil from other non-OPEC countries. During the height of the "squeeze," in late 1973 and into 1974, inventory stocks would have been run down, not built up. On the production side, previously discovered wells, old wells, high-cost stripper wells, and the like, now economically feasible, would have been brought into production; output rates would have increased. In short, there would have been an increase in domestic supply and a significant reduction in import demand, partly because the increased domestic price would also lower demand. All firms would have attempted to avoid dependence upon OPEC oil.

In the longer run, U.S. firms would have undertaken intensive exploration efforts to find lower-cost and politically safer supplies. They would have been willing to use longer-term, higher-cost recovery methods. The number and even size of firms doing exploration and development would have expanded significantly. Alaskan oil would have been shipped to Japan, thereby dramatically decreasing Japan's dependence on OPEC oil.

Conservation efforts would have been stimulated much earlier and with a faster response. As a result, OPEC would have discovered that demand for its higher-priced oil was falling below its anticipated cartel levels, and so the price of OPEC oil would have started to ease—as in fact it did temporarily and to a modest extent, despite the cushioning of the price-rise effect by the U.S. government. Initially, firms would have been reluctant to engage in long-term contracts; the spot price would have begun to fall as OPEC suppliers moved inventories, now temporarily high, into the spot market.

Saudi Arabia and Iran (still working together at that time) would, as the cartel leaders, have had to restrict their own supply more and more to hold up the cartel price. The smaller countries, seeing the decreasing demand for their product, the rising supply from the United States and other non-OPEC members, and the ever-decreasing demand by U.S. consumers, would have recognized very clearly the signs of a falling market. Their response would probably have been to cheat on the cartel by acquiring long-term contracts with "hidden" discounts and by pushing a greater percentage of their output onto the spot market. Iran would probably have been the first OPEC producer to break ranks and abandon the cartel; at that point, the whole structure would have collapsed. As it happened, the fall of the Shah and the collapse of Iranian oil production gave a much needed respite to the cartel—in particular Saudi Arabia, which is now carrying the cartel through limitations on production.

The main lesson in regard to domestic production is that our current oil lack is the result of two decades of declining interest in domestic areas brought on by (1) the lure of cheap foreign sources and (2) inappropriate signals provided by the government. The country is not facing a precipitous decline in potential exploratory areas; the off-shore areas, for example, are 98 per cent untouched, and Alaska has merely been probed. The sense in which the United States has exhausted its cheap and abundant energy supply is simply that the energy, still abundant, is no longer quite so cheap.

PART TWO
Controls and Policy

WHEN A DRAMATIC CHANGE OCCURS in the rate of exchange between the value of a nation's exports and the value of its imports, the government is tempted to legislate something. The temptation is enhanced by the cries of the affected parties urging the government to take action. But to give way to this natural reaction is a mistake.

So far our argument against government involvement in such circumstances has been partial: a case has been made *for* the market solution in preference to centralized and politicized solutions, but the case *against* the political solution has not been presented. This is the main intention of chapter 4.

The method of approach will be to examine some concrete examples in order to lessen the burden of abstract argument. These examples are not to be regarded as merely horror stories of government ineptitude. Even though the governmental agents involved are able and well-intentioned, centralized and politicized decisions often, if not inevitably, lead to such problems as shortages, lines, gluts, and general economic inefficiency because of the manner of decision-making and the kind of information it requires. The comedy of errors to be described in chapter 4 was played out *despite* the good intentions of an army of well-trained and intelligent bureaucrats, not to mention numerous politicians who have proven over many years their political savvy, if not their economic expertise. No matter how well-intentioned, attempts to manipulate market forces usually fail and often make matters worse. Almost from its beginning the oil industry at one level or another has been subject to controls and special legislation. Usually the effect has been to lower the overall efficiency of operation, sometimes by restricting output, at other times by overstimulating exploration, but most often by subsidizing small, high-cost producers at the expense of large and relatively efficient producers.

Notwithstanding this long history of governmental intervention, the 1970s were remarkable for the intensiveness and extensiveness of legislation affecting the oil industry. The important acts in the 1970s include: in 1970, the Economic Stabilization Act; in 1971 and 1973, the four Price Control Phases; also in 1973, the Emergency Petroleum Allocation Act; in 1974, the Federal Energy Administration Act, Energy Supply and Environmental Coordination Act, and Energy Organization Act; in 1975, the extension of the Emergency Petroleum Allocation Act, the Energy Policy and Conservation Act, and the Energy Conservation and Production Act; and in 1977-78, the Department of Energy Organization Act.

Regulations enacted during the 1970s had three main aspects: price controls, allocation and distribution controls, and information collection (even ineffective control requires considerable amounts of information, so that once the decision to control is made, masses of data must be accumulated). While the overall cost of these new regulations, and especially the extra cost passed on to consumers, is hard to measure, it is known to be at least in the hundreds of millions of dollars. Another certainty about controls that we will glimpse in chapter 4 is that controls breed more controls. The point was well put by an employee of the Department of Energy in an unpublished review of regulation:

> Overall, the MOIP [Mandatory Oil Import Program] became burdened with special programs and special exemptions. What started out as a simple program ended up as a complex program with a series of groups with vested interests bringing continual pressure on the MOIP personnel to amend the program in their favor.

Chapter 4 begins with a brief description of the control legislation of the 1970s. It then covers a few major items in more depth in order to give the reader a better sense of what is involved in trying to control a market. The chapter ends with a brief summary of the pervasive influence of governmental involvement and its impact on prices, gasoline lines, and the future supply of oil.

Chapter 5 draws some conclusions and makes some recommendations about policy toward OPEC and toward the oil market.

CHAPTER FOUR

Oil Legislation in the 1970s

PRESIDENT NIXON'S IMPOSITION OF a ninety-day freeze on all prices in August 1971 initiated Phase I of the price controls authorized by the Economic Stabilization Act. The Phase II controls that followed in November were not so temporary and considerably more complex. Phase II was intended to hold price increases to 3 per cent a year, but small businesses, including most oil retailers as well as independent producers and refiners, were exempt from the controls. There were now controls on profit margins. "Cost-justified" price increases required prenotification by the larger firms.

When major refiners sought a price increase in 1972 because their real costs had risen, it was denied. The Price Commission said that an increase could be granted only after a series of public hearings on the whole question of oil prices. The major refiners recognized the pressure and withdrew their request.

The inevitable occurred. During the 1971 price freeze imposed in August, the price of gasoline was fixed at its high summer value and the price of heating oil at its low off-season value. By the winter of 1972-73, heating-oil supplies were too low and gasoline

---

AUTHOR'S NOTE: For much of the material in this chapter I am indebted to two sources: Philip L. Essley, Jr., of the Federal Energy Regulatory Commission, and a report by Paul F. Dickens III, of the Department of Energy's Regulatory and Competitive Analysis Division, entitled, "Effects of Oil Regulation on Prices and Quantities: A Qualitative Analysis" (Reprint No. AR/EI/79-25, U.S. Department of Energy, May 1979).

supplies too high. With the move toward the greater price flexibility of Phase III in January 1973, heating-oil prices (along with crude prices) rose significantly, prompting the Office of Economic Preparedness to "persuade" refiners to increase the output of heating oil at the expense of gasoline. By April 1973 the situation had been reversed: heating-oil supplies were far too high and gasoline supplies too low.

The difficulties were compounded by a halt in refinery construction that began in 1969 and lasted until 1973. This halt was brought about by (1) the Tax Reform Act of 1969, which eliminated the 7 per cent investment tax credit, (2) the difficulty new refineries had in getting access to crude oil under the Mandatory Oil Import Program, and (3) the frozen low refinery margins under Phases I and II. In April 1973, the oil import quotas were abandoned, and Phase III promised independents less price restraint. Within a few months over two million barrels per day of new capacity had been announced.

As a part of this process, the independent refiners (unregulated) increased their demand for oil from independent producers (also unregulated), and the price of oil products rose. As soon as Phase II ended, the majors joined the increased demand for domestic crude. The price rise in March 1973 prompted Special Rule No. 1, which reimposed mandatory controls on the top twenty-four firms (those with sales over $250 million).

Phase IV began in September 1973 by declaring two prices for domestic crude, the "Old Oil" price for oil produced at the same level as in 1972 and the "New Oil" price for oil produced over and above 1972 production. Stripper-well output (less than ten barrels a day per well) was sold at uncontrolled prices. The two-price structure was an attempt to hold down consumer prices without completely stifling the development of new oil areas. Stripper wells were left unregulated mainly for political reasons: compared to the larger companies, stripper-well operators do not produce a lot of oil but do generate a lot of votes.

The market was not allowed to react to the imbalance among the three very different prices created by the setting up of artificial (i.e., noneconomic) categories based on production levels. An unstable situation had been created that could not last.

The other major part of the Phase IV controls froze refiners' and retailers' margins (the difference between the price at which they buy and the price at which they sell, which must include their costs) to prevent retail-price increases during the (government-created) shortages. There was some "leniency" in the controls; refiners and retailers were allowed to pass on their cost increases for some products dollar for dollar. However, three oil products were selected for special controls to hold down the retail price no matter what the cost: gasoline, diesel oil, and home heating oil. (These three products also happen to be the ones used in the calculation of the Consumer Price Index.) The effect was to force a rapid and very large rise in the prices that were allowed to rise to cover increased costs; propane, for example, rose from about fifteen cents per gallon to fifty cents per gallon between 1973 and 1974.

To make price controls work, the government also had to regulate how supplies of the controlled-price products were allocated. And so a new era of controls began in November 1973 with the passage of the Emergency Petroleum Allocation Act and the creation of the Federal Energy Office. Allocation control became a central task of the FEO. (It is now handled by the Department of Energy.)

A Government-Created 'Shortage'

A relatively minor gasoline supply problem that occurred in early 1974 supports my point that regulators usually fail to anticipate the effects of their actions on people's behavior. On February 12, 1974, the newly created Federal Energy Office announced it was increasing the fixed service-station margin by two cents per gallon. However, Cost of Living Council rules prevented the increase from taking effect until March 1.

The result was that service-station sales fell immediately; sales were limited to a small maximum per customer; stations closed early and opened late or not at all; supply lines and storage tanks were backed up; and a survey of the major marketers of gasoline showed that their own station operators were refusing deliveries.

Long gasoline lines developed. Within a few days after the two-cent margin increase went into effect on March 1, everything returned to normal; the lines disappeared. This is our first example of "mandated" gasoline lines: although the lines were definitely not the intention of FEO efforts, they certainly were the effect.

Service-station operators, even those selling a major's name-brand product, are predominantly independent members of a highly competitive business. Service stations are numerous and small; the cost of opening a station is low, and turnover is extraordinarily high. But—and this is one of the fundamental points made in the first chapter of this study—government regulation can produce the effects of a cartel where they would otherwise be impossible.

Consider the situation described above: as of February 12 margins were well below market level; a major margin increase was announced on February 12, but it was not to take place until two and a half weeks later. The inevitable result was that each operator acting individually saw that it was in his interest to hold off on his sales until the new margin took effect. For this to work he had to be sure that he could indeed sell all he wanted at the higher margin; this was guaranteed by the restricted supply situation and the previous three years of margin restraint, coupled with substantial inflation. There was no collusion among dealers. By the stroke of a pen, the FEO produced a "shortage."

One of the roles played by market prices is the allocation of the available supply among various classes of consumers. As soon as the price is held down below a market-clearing level, the allocation of the available supply—or, to put it differently, the allocation of the burden of the shortage—is politicized. Formulas for who gets how much and when must be decided. If these decisions are in the hands of government, consumer groups will form to apply political pressure in order to be declared a favored supply group.

As a result of price controls coupled with a big increase in demand for low-sulphur oil stimulated by environmental concerns, there developed in late 1972 and early 1973 an increasingly severe scarcity of domestic low-sulphur fuel oil. Producers with access to low-sulphur oil kept it for their own refineries or for refiners with whom they had long-term contracts. The inevitable result was that many independent refiners who had been buying on the spot market or who had only short-term contracts were unable to get enough environmentally acceptable crude oil. Part of the difficulty faced by the FEO was that the supplies of many products, especially gasoline, were below normal. And so a large number of independent refiners, marketers, and service-station operators appealed to the government to supplant the market and legislate to them supplies of low-sulphur crude. A series of mandatory allocation schemes was the next step. We will look at some allocation controls later in this chapter.

The post-1975 period culminated in the creation of the Department of Energy, now a multi-billion-dollar operation—justified perhaps on the grounds that a big industry requires a big agency to supervise it. The latter half of the decade was marked by a substantial increase in the information requirements of the government and in both the extensiveness and intensiveness of control, despite the often repeated claims that the new bureaucracy was of a "temporary nature" and that there would be a "planned phasing out of controls." Recall that the process began with a ninety-day freeze in 1971; then the Federal Energy Administration was created merely to handle the temporary crisis of the 1973 embargo; its successor, the Federal Energy Office, was supposed to be phased out as well, but instead grew and was transformed into the Department of Energy.

The major crude-oil price-control change after 1976 established three major prices: (1) Lower Tier (similar to Old Oil prices); (2) Upper Tier (similar to New Oil prices); and (3) Stripper and certain types of Tertiary Recovery production, which could be sold at the world price. In addition, there were separate prices for Alaskan North Slope and Naval Petroleum Reserve oil. From an efficiency point in view, without allowing for quality differences and transportation costs, there should be only *one* price of oil, not five.

The total 1979 estimated governmental expenditure on energy, not including nuclear, was $7.1 billion or about $1.78 per barrel of oil produced, but this figure includes research and other items besides the cost of regulation. Establishing the "real cost" of the controls alone is difficult. Analysts Arrow and Kalt* produced some extremely conservative estimates (conservative by design, since Arrow and Kalt chose—quite rightly—to produce a "reasonable" lower bound on the cost). The estimate is $2.5 billion per year. Part of the difficulty in calculating the real cost is that controls always benefit somebody, and their gains must be subtracted from the losses suffered by others.

Arrow and Kalt ask about the *gains from decontrol,* rather than the *costs of imposing control*, which is certainly the prior question. They conclude: "It is possible to make everyone better off by decontrolling oil prices and paying sufficient compensation to those who would face higher prices" (p. 34). However, this statement neglects to mention those who lost when the controls were initially imposed—lost heavily and were *not* compensated.

Three short case histories of controls and their effects—the entitlements program, the role of distribution controls and mandatory set-asides in the shortages observed in the spring of 1979, and the "Windfall Profits Tax"—will help to illustrate our points.

The Entitlements Program

The entitlements program was set up to support a system of multiple prices for crude oil. Politically, it was regarded as "fair" to tax those refiners who have access to cheap domestic oil in order to compensate those who have to use expensive foreign oil. The effective outcome, no matter what the intention, is to tax domestic producers and subsidize foreign suppliers. This kind of taxing makes little economic sense, especially when the subsidized foreign suppliers have formed a cartel.

Under the entitlements program, a refiner who has access to lower-priced domestic oil must purchase, from a refiner who uses higher-priced foreign oil, "entitlements" to (i.e., rights to buy) that domestic oil. This equalization measure raises the costs of the

---

*Kenneth J. Arrow and Joseph P. Kalt, *Petroleum Price Regulation: Should We Decontrol?* (Washington, D.C.: American Enterprise Institute, 1979).

domestic-oil refiner in order to compensate the foreign-oil refiner for his higher costs.

The total number of entitlements issued each month is equal to the total number of barrels of Old Oil (i.e., oil from leases in production in 1972) produced. (I am using the rules of the earlier and simpler Emergency Petroleum Allocation Act rather than the more complex rules of the Energy Policy and Conservation Act to explain the principle of entitlements.) The number of entitlements available to any one refinery is determined by multiplying the number of barrels it refined in a particular month by the percentage of Old Oil in the total oil refined industry-wide during that month. For example, if refinery A processed 100,000 barrels in a month, and half of the total oil refined in the industry that month was Old Oil, then refinery A would receive 50,000 entitlements, i.e., the right to buy 50,000 barrels of Old Oil. Now, if refinery A had in fact purchased only 30,000 Old Oil barrels, it would have 20,000 entitlements left over to sell to someone else. Suppose refinery B and refinery C processed 20,000 barrels of Old Oil each but have only 10,000 entitlements each (because each processed a total of 20,000 barrels in a month when the industry-wide Old-Oil-to-total-oil figure was 50 per cent). B and C are each short 10,000 entitlements, and so each buys 10,000 of the 20,000 entitlements refinery A has for sale.

The price for an entitlement is the difference between the per-barrel price of foreign oil and the per-barrel price of Old Oil; for example, if foreign oil is selling at $15 and Old Oil at $5, then an entitlement is worth $10. To see the effect of entitlement, compare two refineries: one bought 100 barrels of Old Oil at $5, the other 100 barrels of foreign oil at $15. Each had a total output of 100 barrels, and so, with an industry-wide 50 per cent ratio of Old Oil to total oil, each gets 50 entitlements. The refiner of domestic oil must therefore buy 50 entitlements from the importing refiner. This means, then, that the post-entitlement per-barrel price for the Old Oil refiner is $10 (oil cost plus entitlement) and for the foreign-oil refiner is also $10 (oil cost minus income from sale of entitlement).

Consequently, there is no price difference to the refiner between domestic oil at $5 and foreign oil at $15. But since domestic oil is restricted in supply but foreign oil is not, refiners buy much more imported oil than they would if they had to pay the higher import price without compensation through entitlements.

There are further problems. Because the cost of imported oil has been lowered by entitlements, the final price of the products of oil refining will be less. More will be bought, therefore, and once again imports will increase. An equally serious effect is that domestic supply is artificially restricted because the domestic *producer* is still getting only $5 per barrel. The additional $5 paid by the refiner who purchases domestic oil is a subsidy of the importer of foreign oil. But the subsidy does not end with the importer; its value gets transmitted back to the foreign-oil producer through increased consumption of his oil. The ultimate indirect effect, then, is to tax domestic *producers* of oil and transfer that tax as a subsidy to foreign producers.

The Market Alternative

To see more clearly what happens in this manipulation of the market, contrast the situation as it was with what it might have been had the government not intervened. The high-cost foreign oil needed to meet the gap between domestic demand and domestic supply initially determines the market price at $15 per barrel, but *every* producer of oil, domestic or foreign, gets $15. Because the consumer price is higher, consumption falls relative to the pre-OPEC price situation and is lower than in the entitlements scenario. The volume of imports is lower also.

The new higher price of oil means domestic producers get an immediate onetime increase in their wealth; the total amount of increase depends on how much oil they are holding in inventory at the time the price goes up. For example, if the XYZ Oil Company has *holdings* of 1,000 barrels of oil and the price has jumped by $10 per barrel, the wealth of the owners of XYZ has increased by $10,000. Secondly, XYZ can now get a new higher price for each new barrel of oil it produces. It can therefore increase what it spends on exploration and production up to a new higher cost of

$15 per barrel—a threefold increase in the feasible costs of finding and producing new oil. As domestic output responds, imports fall yet again. Increased domestic producer revenues and the higher expected price serve to increase domestic employment and significantly raise the Treasury's tax receipts.

The overall short-run outcome of this alternative of no government intervention is that, relative to the situation under the entitlements program, consumers pay more but consume less, the United States produces more and imports much less, we do not subsidize foreign suppliers out of domestic consumer expenditures, and tax revenues are up. These are the responses that would occur within the first year or two.

The longer-run results—again relative to the results of government intervention—would include higher domestic output, higher consumption at lower prices, much lower imports, and higher tax revenues. And, comparing the long-run results to the situation before the OPEC-instigated price increases, we would find higher domestic output, lower consumption at higher prices, and higher tax revenues.

Since both the short-run and the long-run response seem to be so much superior to the situation under the entitlements program, the reader may be suspicious. What's wrong with our analysis?

As it stands, nothing; but it does not include the whole picture. Oil prices and profits are regarded by politicians as politically sensitive, especially in a period of high inflation. And there would be no escaping the fact that initially *domestic oil producers*, whether majors or not, would receive a spectacular bonanza.

Here then is the rub. The initial working of the market would produce higher domestic prices and enormous gains to domestic producers. Even though in the long run competition would bring profits back to normal levels, output would rise, price and imports would fall, and oil security would be increased, nevertheless the initial phase is politically unacceptable. A sudden large increase in price followed by mammoth profits would demand political intervention. The entitlements program was part of the response to this political problem, understandable but costly.

There were, of course, some exceptions that made the entitlements program more complex than my simple summary of it suggests. One of these special provisions was the Small Refinery Bias. The bigger a refinery is and the more it is used, the more efficient it is. A large refinery operated at near capacity will waste a much smaller percentage of crude oil and will produce a greater variety of useful products than a small refinery. Nevertheless, in order to be "fair" in the entitlements program, the government gave special treatment to small refineries. In October 1977, about 5 per cent of the total entitlements were for small refineries (those that produced less than 175,000 barrels per month); but by 1979 the figure had risen to 13 per cent. The larger a small refinery was, the less its Small Refinery benefit; in other words, the more inefficient the refinery, the greater the subsidy. One result of the Small Refinery Bias was a boom in the construction of small refineries. This was, then, another case of wasting crude oil through regulation that encouraged inefficiency.

Gasoline Lines by Fiat

Many of us found ourselves waiting in lines to buy gasoline during the late spring and early summer of 1979. What was the cause of these long, annoying waits?

First, the flexibility of oil firms to redirect gasoline deliveries where they were needed as evidenced by changes in relative prices had been severely reduced by the joint effect of price controls and allocation controls. The allocation controls gave each state 5 per cent of a base-period demand to be allocated at the state's own discretion; this is what is called the "5 per cent set-aside." Certain categories of use received 100 per cent of their current demand; in the controls that became effective in May 1979 they were:

industrial use	telecommunications services
commercial use	passenger transportation
social agency use	services
Department of Defense	aviation ground support
agricultural production	equipment
energy production	freight and mail hauling by
sanitation services	truck

Any quantity adjustment caused by supply disruptions is therefore concentrated on the private-automobile demand.

In the spring of 1979, the Department of Energy took action in response to criticisms that the gasoline allocations made no allowance for unusual growth. The regulations were altered to allow a buyer—say a service-station operator—to choose as his allocation base either his actual purchases during the specified 1978 base period or his average monthly purchase for a five-month period during the 1978-79 winter. The alteration was well-intended and, as written, sounds like a good idea. It took effect at the beginning of May 1979; the date is significant.

In the summer of 1979 most of the majors were able to deliver about 95 per cent of the previous summer's allocation. But the effect of the controls was that rural dealers obtained about 101 per cent of their prior year's amount and urban dealers only about 91 per cent, with many dealers getting only about 82 per cent—instant misallocation, too much gas in the country and too little in the cities.

During this period several state governors asked the oil companies to exercise the option they had under DOE rules to increase the total allocation to an area of shortage by up to 5 per cent without prior DOE approval; this action, however, had to be reported to the DOE for its surveillance. But with the total supply at controlled prices below demand, the question was from whom to take in order to give more to those who had a shortage. Given the existing regulatory climate and the state of public opinion, it is no great wonder that the oil companies—particularly the majors— would have nothing to do with such reallocations on their own; since the DOE was deciding how much each area got, let the DOE solve the problem it had created. For an oil company to move oil from one state to another under such circumstances would have virtually guaranteed the filing of suits by every state, country, or municipality from which even a tiny percentage of its supply was withdrawn.

This example illustrates the difficulties inherent in centralized control and attempts to circumvent the market. The frustration of the summer of 1979 was certainly not the intention of the DOE policy implemented in May 1979. But bureaucratized and politicized allocation schemes for goods used privately are poor substitutes for markets, even imperfect markets.

A similar problem created by attempts at centralized control was reported by *Time* magazine in May 1980. Standard Oil of Ohio (Sohio) was one of the companies that developed the Alaskan North Slope, and under the DOE's pricing structure the price of Alaskan oil was $17.88 per barrel. Sohio had access to oil at that price, therefore, when the average price paid by other companies in its area was $24.81. This meant that Sohio service-station operators could buy at a lower price than their competitors. But since DOE regulations also limit dealer margins, the only course open to Sohio dealers was to lower the price of gasoline, which they did by ten cents per gallon. Sohio pumps were immediately swamped with eager customers. That was Act I. Act II: A newly formed coalition of those without access to Alaskan oil called the Ohio Independents for Survival complained to the DOE. The DOE ordered Sohio to raise its prices in order to match independents' prices. Act III: The White House worried that ordering Sohio pump prices to be raised might have a bad political effect on the upcoming Ohio presidential primary. The DOE then reversed itself.

Under this kind of regulation, supply decisions have been politicized, economic efficiency has become a minor concern, and, despite the DOE's concern about equity, perceived inequities multiply rapidly.

"Windfall Profits Tax"—An Oil Discovery Excise Tax

The first point of interest about the Windfall Profits Tax is that it has absolutely nothing to do with "windfall profits," which are *unanticipated* earnings caused by unforeseen shifts in demand or supply. The gain occurs only once. Proponents of the tax are correct in stating that a tax on *past* profits does not affect *current* output. But the corresponding shift in industry expectations about taxes on *future* gains leads to a reduction in exploration.

Take the case of a wildcat firm that stays afloat on a small stream of oil discoveries, say about one in five attempts. The firm survives but does not grow; no one is about to get rich on the returns, which are less than the owners could make by getting out of the business and selling shoes. What keeps a wildcatter in the oil game is the chance that someday he will hit a big reservoir—a very big reservoir. Suppose now that the wildcatter is told that if his big chance does come and he strikes a big deposit, government will tax the return heavily as a windfall profit. That prospect lessens the expected return over the long haul, and the oil exploration business is now more risky than before. At best, firms will be more cautious in exploring and will be less inclined to take the riskier gambles; at worst, small firms will drop out of the business and bigger firms will shrink their exploration activities.

The Windfall Profits Tax is not a tax on windfall profits; it would more appropriately be called the "Oil Excise Tax." It sets up three major tax categories corresponding to Lower Tier (Old Oil) and Upper Tier (New Oil) plus a new category, Newly Discovered Oil. (Other categories are treated in a similar manner to one or another of these three.)

The first thing to note is that it is oil *revenues,* not oil *profits*—accounting or otherwise—that are taxed. All refineries pay the OPEC-determined world oil price for their crude supply, but that is not what the producers of the crude receive.

For oil at the lowest price under the price-control system, Lower Tier or Old Oil, the government collects as a tax between one-half and three-quarters of the difference between the world price and the base price of $6 per barrel. There is provision neither for a price increase to cover an increase in the real costs of production nor for the fact that revenues from producing fields help to finance exploration.

Upper Tier or New Oil, that which was discovered after 1972, is taxed similarly, with the base price being $13. In addition, oil companies cannot assume that future gains in New Oil revenue will be taxed at the same rate; especially large future gains are very likely to be taxed at a higher rate. Normally an oil company counts on such large gains to offset its exploration losses and disappointingly small gains. To tax large gains at a higher rate lowers the long-run overall expected gain for the firm. This decreases the firm's willingness to engage in risky exploration and development, including new methods of recovery.

The third category includes Newly Discovered (since 1978) Oil and Tertiary Oil (oil recovered by advanced and more expensive techniques). Once again the concept is that the maximum marginal price needed to discover oil is halfway between the $16 base price of this type of oil and the world (OPEC) price, and so the other half of that price difference goes to the government as a tax. This taxing structure makes no allowance for a rise in the real costs of exploration and development, to mention just one problem, and, as we have seen, those costs have risen dramatically during the last few years.

In the long run the effect of the tax is to lower the amount of oil supplied. By reducing the wealth of those engaged in oil exploration, it lowers both the extent and the type of exploration that is undertaken. The result: ever-expanding imports.

A more accurate description of the "Windfall Profits Tax" is "The Government's Windfall Tax Gain." The addition to governmental revenues anticipated when the tax was enacted was $227.7 billion. From the government's point of view, the Windfall Profits Tax is an unbeatable source of ever-expanding revenues; as the world price of oil rises, tax receipts rise and we can pretend that the oil firms rather than the consumer ultimately bear the tax.

Lessons About the Political Control of Oil

The examples of government regulation that we have been examining offer a number of important lessons for an observer of the oil industry or, in fact, of any industry. But before we look at these lessons, let us clear away some possible unintended implications. I did not intend to suggest that government is particularly inept; one could as easily present horror stories of business ineptitude. Nor did I wish to imply that people in government, either legislators or bureaucrats, are especially likely to make mistakes. All human beings make mistakes; I see no reason why politicians or bureaucrats should be any more mistake-prone than any other group of human beings, including businessmen.

What I do mean to say, however, is that some systems of decision-making are more suitable than others for certain tasks. Government must "govern," but markets are superior for getting economic tasks done.

Our first lesson is that while the oil companies are the *visible* sources of our frustration, the causes lie deeper. We have seen that the spot shortages, wide price variations among firms, gasoline lines, and even the decline in service are natural outcomes of government efforts to control oil prices and then to allocate gasoline in a vain effort to make the price controls feasible.

The initial steep increase in the price of gasoline and heating oil was a direct effect of OPEC's actions, and we would like to think of OPEC members as the villains. But they are merely looking after their own nations' interests, having felt "exploited" by the West for decades. OPEC nations are doing what any other nation would be likely to do in the same situation: turning the terms of trade in their favor.

Is our government, then, the villain? No. It is the source of much of the difficulty, but unintentionally so. Much of the oil problem stems from the government's efforts to subsidize the oil consumer by taxing the domestic producer, to shield the voter from the inevitable increase in the real cost of oil and its products, and to implement politically popular goals such as those prescribed by the Environmental Protection Agency (many of whose actions led to higher consumption and lower domestic production).

I am convinced that an unimpeded and *unsubsidized* oil market, despite its imperfections, would be far more effective in obtaining, producing, and distributing oil and its products than a Department of Energy even twice its current size. I believe that the market, which reflects the net result of the decisions of all who participate in it, does a far better job of allocation than schemes of centralized control. With centralized schemes, mistakes are more serious because they are more general and usually market-wide; they are also more numerous because the "error-prone" are not weeded out by competitive pressure. While individual decisions in a market are localized and require little information, bureaucratic procedures require enormous amounts of data. This vast amount of information is extremely costly to acquire, and much of what is needed is unavailable.

Market solutions are flexible; they continually and quickly adapt to new circumstances. And by providing rewards to innovators, they encourage the anticipation of new events. Bureaucratic procedures are slow and discontinuous, and they react retroactively to changing circumstances.

While many observers would grant that the market is more efficient than systems of centralized control, they nevertheless reject it as unfair. In an uncontrolled market, they say, the relatively rich owners of oil stocks would gain and the relatively poor consumers of oil products would lose; controls are needed, therefore, to tax the rich and subsidize the poor.

The first response to this is that if the concern is for equity in income received, then controlling price and allocation to achieve a more "equitable" state is not only inefficient but also usually counterproductive. We already have numerous procedures for redistributing income; we need not reduce productive efficiency by adding yet another. Let us leave income redistribution to the income tax.

Secondly, the correlation of "the rich" with oil producers and "the poor" with oil consumers is grossly overdrawn. A careful analysis of the income redistributions involved might well show that on balance the richer gained relative to the poorer, but one would need some sophisticated and very detailed analysis to show that result.

Earlier I tried to show that the rich-poor distribution is overdrawn in another sense. We are *all* poorer by virtue of the fact that foreign oil is now more expensive. But domestic oil producers and all their employees and owners are better off, not because they have exploited the consumer, but because they are now engaged in the production of a more valuable commodity. As a society we will all gain if more people move out of their existing occupations and into oil exploration and production. However, this will happen in a politically free society only if people have an incentive to move. This is exactly what the market reaction provides.

Why Governments Intervene in Markets

If it is as counterproductive as I suggest for governments to intervene in markets like this, why do almost all governments do it? The answer would be the subject for another book. But perhaps we can get a brief glimpse of it. As we try to do so we need to keep three points in mind: (1) the arguments I have presented are almost entirely economic, not political; (2) the politically expedient and an increase in the general welfare are not inevitably compatible, and indeed, are often antithetical; (3) whenever a big economic problem arises, the public expects the government to "do something," at least to pass some legislation.

We have seen that throughout industry various coalitions or groups organize to ask the government to protect their common interest from an economic threat. When domestic oil producers during the 1950s urged the government to keep out "cheap foreign oil"—namely, Middle Eastern oil—they were acting just as the manufacturers of shoes or TV sets or the growers of oranges might act. In all cases the directly affected parties are seeking a "subsidy" of some form from the government.

The government can grant this only at the expense of the rest of society. If Florida fruit growers benefit by limitations on the importation of Mexican fruit, the rest of us bear the cost in higher prices and restricted options. If rent control is imposed, landlords are implicitly taxed. If the price of oil is held below its market price, consumers benefit in the short run at the expense of oil producers and distributors. Who gets what gain depends on the balance of political power; most simply, if many can be benefited at the cost of a few or if a few can be handsomely benefited at the modest cost of many, the political trade will be consummated—even if on purely economic grounds the overall result is a loss.

Another major stimulus to governmental involvement is that it is expected, if not overtly requested, by the public. A consumer's natural *political* reaction—as distinguished from his *economic* reaction—to a problem such as a disruption in supply is to demand that the government do something to "guarantee" delivery of the product at its *current* price. People do not like to change; they prefer to pass on to someone else the burden of adjustment to new conditions. More thoughtful consumers still demand some political action, since that indicates to them a positive response by the government. Allowing the market to work itself out appears to be ignoring the problem; inaction has the unattractive appearance of seeming to "allow events to dictate" our actions.

These are some of the forces behind the demand that the government play a role in the allocation of scarce and widely used resources. It is no wonder, then, that government intervenes.

CHAPTER FIVE

Policy Implications and Positive Proposals

NOW WE MUST TRY TO pull together the information, ideas, and insights we have acquired and answer our original questions. A major question was about the prices of gasoline and heating oil: Why are they as high as they are, and will they continue to increase? The first part of the question can be answered; about the second we can only speculate.

The current price level is due in part to OPEC, in part to our own inflation and the consequent devaluation of the dollar, and in part to our government's misguided actions. The major price rises in late 1973 and 1979 were clearly OPEC-inspired; the first increase was unmistakably large and effective, the second relatively muted. A domestic inflation rate that has approached 15 per cent a year has certainly contributed substantially to the overall increase in pump prices. The third factor, U.S. policy, subsidized imports, raised world demand for OPEC oil, and thereby helped to consolidate the cartel.

A number of U.S. actions contributed to this extraordinary turn of events. First, the government's "entitlements" program, in which it attempted to hold down the domestic price of oil by indirectly taxing domestic producers, led to a substantial increase in imports by stimulating domestic demand and lowering the incentives to domestic supply. Other domestic policies, such as regulations by the Environmental Protection Agency, however well intended, led inevitably to substantial increases in demand at all prices. Their net effect was to raise domestic and world demand for OPEC oil, a turn of events much resented by our European friends.

Whether or not oil prices continue to rise faster than the general rate of inflation depends on several factors. A shift in regulations that will ensure that U.S. consumers pay the world price will produce a sizable, but one-time only, increase in price. This will further restrict demand but stimulate domestic supply. Offsetting this effect, however, is the excise taxation of producers through the so-called Windfall Profits Tax, which helps to inhibit domestic supply and thereby to keep prices somewhat higher than they otherwise would be.

On the international front the signs are mixed. As European countries and Japan succeed in their efforts to raise oil productivity and discover oil substitutes they will need less oil to support any given level of Gross National Product. And Third World countries are making increasingly successful attempts to lower their dependence on foreign oil.

At a time when there were most encouraging signs of high tension and conflicting interests within OPEC, the cartel was rescued by the outbreak of the Iraqi-Iranian war. The cartel had previously had a similar rescue through the dramatic decline in Iranian output after the departure of the Shah.

The OPEC cartel is dominated by its Arab members, and the dominant Arab country is Saudi Arabia. Because OPEC is composed of nation-states rather than private firms, the classic notions of cartel generation must be modified to take into account political forces. This political dimension gives the cartel both strength and weakness. The basic internal economic forces for dissolution of the cartel through cheating by the smaller members are still there, but these forces are counterbalanced by the political pressure that Saudi Arabia can muster and the adherence of cartel members to "Arab unity." If that sense of unity were to fail (and there has been some weakening), then OPEC would collapse quite quickly.

The development of the international spot market and the efforts by OPEC members to integrate their operations down to the retail level are good news. Both actions will weaken the ability of the leading OPEC producers, Saudi Arabia in particular, to police the oil prices charged by OPEC members. This weakening coupled with the enormous gains to be made by cheating on the cartel agreement will lead eventually to the cartel's demise.

Another encouraging development is the massive worldwide exploration effort now under way. This will begin to produce ever-increasing amounts of oil within a few years (there is about a ten-year gap between the initiation of oil search and the resultant flow of oil through the pipelines). New oil sources will eventually bring considerable downward pressure on prices, and this may offset U.S. actions that cause upward pressure.

What Caused Those Gas Lines?

The next question: What caused those gasoline lines in 1974 and 1979? At the time an angry public directed its blame at the biggest target, the majors. But we now recognize that the lines in the spring of 1979 were caused by the natural—if self-serving—reactions of thousands of independent retailers to changes in DOE regulations. And the lines in the winter of 1974 were caused by misguided government attempts to build new inventory stocks in oil and to hold down prices. At the most charitable level, one must say that the timing of such a decision to build emergency stocks—especially since it was coupled with price controls—was wrong.

The gasoline lines, in other words, are merely the outward and visible signs of an inward bureaucratic malaise. The treacherous path to apparently stupid blunders in oil-product allocations begins innocently enough with the political dictate to set varying prices for the sake of "equity" and to lower the average level for the sake of the "poor"; the regulatory agency attempts to do this without knowing exactly what the inequities are and precisely how the poor are affected. To make these non-market-related prices stick, the agency gets itself embroiled in more and more complex schemes of quantity allocation.

Before we cast too many aspersions on the unfortunate bureaucrats charged with supervising this mess, we should recognize that few if any of us could do any better. The problem lies in the difficulty of bureaucratic control, not in the capabilities and intentions of the controllers. Market conditions change continually and in subtle ways; bureaucracy cannot keep pace and could not gather enough information in any case. The outcome, however undesirable, is the fault not of their methods but of their mission.

Most Americans have been convinced that the oil companies have taken advantage of them—"ripped them off"—during the past decade. I have tried to show that this conviction is wrong, that the oil companies, majors and independents alike, are not themselves responsible for our current problems. I do not mean to ascribe to oil companies any particular altruistic qualities but merely to recognize that they do not play the central role in the drama.

In examining profits, we saw that until very recently oil company profits, though volatile, were on the average modest, comparable to the general level of manufacturing returns. Also, we saw that even if all oil-company profits were eliminated, the immediate gain to the consumer would be from 3 to 5 per cent of the gasoline pump price—not an amount to generate much excitement.

We also looked into the matter of where oil profits go: are oil companies investing in oil exploration or in such unrelated enterprises as department stores and almond orchards? We saw that the charge of not investing in oil is largely an empty one. If the oil companies see their oil profits rise, they will invest in oil to try to reap an even larger return and stimulate their firms' growth.

Further, we saw that the industry is competitive at all stages of production and distribution. But we also noted the inclination of the oil industry, like any other, to seek government aid and subsidies whenever the opportunity arises. However, we should not say that "the industry" seeks political largess; it is particular groups within the industry that do so, and what benefits one may hurt another. One constant theme of government involvement in an industry is the support for "small business" rather than the leaders in the industry.

How to React to OPEC

A major development during the past decade is that the United States is no longer the dominant supplier in the world market and must take into account the role of OPEC. How should it react to OPEC actions?

OPEC is an effective cartel, and the only effective way to deal with an international cartel is through economic forces. One avenue of attack is to increase competing supplies. The U.S. government can do this at home by removing price and allocation controls and differential tax rates that either impede the efficiency of production or raise the risk of exploration. It can also encourage non-domestic, non-OPEC supplies by reducing the risks, costs, and taxes faced by U.S. firms operating abroad. Long-term contracts for buying oil from other countries should be based on economic grounds, not political expediency. Mexico is a good example of our misused opportunities in this regard.

The second avenue of attack is to restrict demand, but here we must weigh the gains from the restriction of imports against the economic losses from reduced fuel use. When the market is permitted to work normally, it achieves the appropriate balance between the benefits of import reduction and the benefits of fuel use.

The third avenue of economic attack on the OPEC cartel is to pursue a strategy that makes the policing of the cartel difficult and that capitalizes on the inevitable differences among the members. Decentralized decision-making is the key. If the whole of OPEC were to deal exclusively with the U.S. government, its power of survival would be considerably enhanced. The more numerous the contracts and the more varied they are in nature, timing, size, and duration, the more difficult it is for the OPEC leader, Saudi Arabia, to police the output rates of the cartel members. The cartel can survive only if Saudi Arabia can assure restricted output levels by the other members. The buildup in the relative importance of the spot market will contribute substantially to the breakup of the cartel, because it provides an outlet for oil less easily policed than the usual long-term contracts. When demand decreases, spot prices fall, and those supply countries relying most heavily on the spot market for selling their oil experience even greater pressure to renege on the cartel agreement to maintain low output rates.

A final question: Do we really need the Department of Energy? My answer is that we do not, never did, and will not in the foreseeable future need a Department of Energy to supplant the normal workings of the market. (Other DOE activities such as collecting and processing data and sponsoring research are another matter.) Left to itself, the oil market would function as smoothly as ever and would enhance the probability of the demise of the OPEC cartel. Prices would be higher in real terms than they were in the 1960s. However, the total real cost of oil consumption to society as a whole would be less in an unimpeded market system than in the current bureaucratic morass.

Courtesy of John Trever, *Albuquerque Journal.*

CONCENTRATION RATIOS IN MAJOR SECTORS
OF THE OIL INDUSTRY

SECTOR AND YEAR	PERCENTAGE OF TOTAL INDUSTRY SALES	
	Top 4 Firms	Top 8 Firms
Crude oil reserves		
1970	37.2	63.9
Crude oil production		
1955	18.8	31.1
1970	30.5	50.1
1972	29.4	46.9
Total crude oil and natural gas liquids production		
1972	28.8	45.8
1973	26.2	42.1
Refining capacity		
1972	33.1	59.0
Refinery crude throughput		
1974	31.5	52.8
Petroleum refining		
1955	32.8	57.5
1972	31.0	56.0
Interstate petroleum pipelines, by volume		
1974	30.9	52.5
Non-Communist international tanker market		
1972	14.3	17.5
Gasoline sales		
1954	31.2	54.0
1972	29.0	51.6
1973	31.0	54.0
Natural gas sales (interstate)		
1955	23.0	35.0
1971	25.3	42.8
Lubricating oils and greases		
1967	38.0	50.0
All U.S. manufacturing, average		
1970	40.1	60.0

SOURCE: Data from *The Structure and Competitive Behavior of the Petroleum Industry—A Fact Sheet* (Washington: Federal Energy Administration, April 1976) and William A. Johnson et al., *Competition in the Oil Industry* (Washington: George Washington University, 1976).

Appendix 2

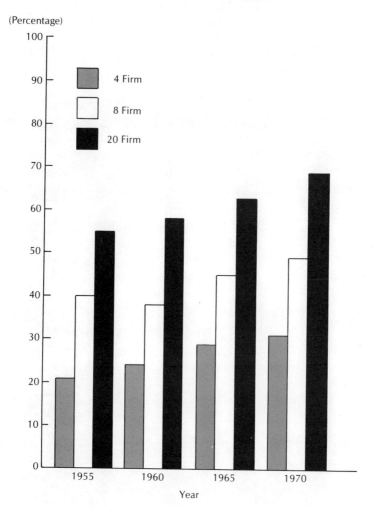

CONCENTRATION RATIOS IN
CRUDE OIL PRODUCTION

SOURCE: *The Structure and Competitive Behavior of the Petroleum Industry —A Fact Sheet* (Washington: Federal Energy Administration, April 1976).

CONCENTRATION RATIOS IN
SELECTED MANUFACTURING INDUSTRIES

INDUSTRY AND YEAR	VALUE, 1970 SHIPMENTS (billions)	NO. OF FIRMS	PERCENTAGE OF VALUE OF SHIPMENTS		
			Top 4	Top 8	Top 20
Motor vehicles	$28.2				
1967		107	92	98	99+
Blast furnaces					
and steel mills	21.5				
1954		. . .	55	71	86
1967		200	48	66	83
Electronic computer					
equipment	5.7				
1967		134	66	83	92
Construction machinery	4.8				
1963		561	42	53	70
1967		578	41	53	72
Tires and inner tubes	4.6				
1963		105	70	89	97
1967		119	70	88	97
Plastics	4.4				
1954		149	47	69	88
1967		508	27	43	64
Metal cans	3.9				
1954		109	80	88	96
1967		96	73	84	94
Tobacco	3.5				
1954		12	82	99+	
1967		8	81	100	
Aluminum rolling	3.5				
1963		166	68	79	89
1967		155	65	79	89
Average all U.S.					
manufacturing					
1958			37		
1970			40		

SOURCE: Date from *The Structure and Competitive Behavior of the Petroleum Industry — A Fact Sheet* (Washington: Federal Energy Administration, April 1976).

Appendix 4

CONCENTRATION RATIOS FOR
MAJOR U.S. INDUSTRIAL SECTORS

SECTOR	PERCENTAGE OF VALUE OF SHIPMENTS			
	Top 4 Firms		Top 8 Firms	
	1967	1972	1967	1972
Primary aluminum	W	79	100	92
Flat glass	94	92	98	W
Motor vehicles	92	93	98	99
Primary copper	77	72	98	100
Tires and inner tubes	70	73	88	90
Aircraft	69	66	89	86
Industrial gases	67	65	84	81
Alkalines and chlorine	63	72	88	91
Synthetic rubber	61	62	82	81
Blast furnaces and steel mills	48	44	66	65
Industrial trucks and tractors	48	50	62	66
Semiconductors	47	57	65	70
Weaving mills (synthetic)	46	39	54	54
Ship building and repairing	42	47	59	63
Construction machinery	41	43	53	54
Lubricating oils and greases	38	31	50	44
Fertilizers	35	35	55	53
Petroleum refining	33	31	57	56
Weaving mills (cotton)	30	31	48	48

W: The government withholds these data to avoid disclosing information about individual companies.

SOURCE: William A. Johnson et al., Competition in the Oil Industry (Washington: George Washington University, 1976), p. 4.   Source cited there: 1967 and 1972 Census of Manufacturers: Concentration Ratios in Manufacturing (Washington: Department of Commerce, Bureau of the Census).

Appendix 5

Concentration Ratios in Other U.S. Industrial Sectors

Sector	Percentage of Value of Shipments			
	Top 4 Firms		Top 8 Firms	
	1967	1972	1967	1972
Electric tubes (receiving)	94	95	99	99
Electric lamps	91	90	95	94
Hard surface floor coverings	89	90	99	98
Turbines and turbine generators	76	90	82	96
Chewing gum	86	87	96	98
Primary batteries	85	92	95	97
Cathode ray picture tubes	84	83	98	97
Cigarettes	81	84	100	100
Typewriters	81	W	99	W
Sewing machines	81	84	92	92
Gypsum products	80	80	93	93
Chocolate and cocoa products	77	74	89	88
Household vacuum cleaners	76	75	94	91
Woven carpets and rugs	76	78	93	91
Electrometallurgical products	74	74	90	90
Medicinals and botanicals	74	59	81	75
Household refrigerators and freezers	73	85	94	98
Metal cans	73	66	84	79
Mineral wool	71	75	84	89
Electron tubes (transmitting)	70	55	87	80
Soap and other detergents	70	62	78	74
Photographic equipment and supplies	69	74	81	85
Cutlery	69	55	77	67
Explosives	67	67	91	86
Greeting card publishing	67	70	79	78
Beet sugar	66	66	96	96
Transformers	65	59	78	75
Thread mills	62	62	81	77
X-ray apparatus and tubes	62	54	77	75
Storage batteries	61	57	83	85
Glass containers	60	55	75	76
Primary zinc	59	66	90	W
Phonograph records	58	48	67	61
Soybean oil mills	55	54	76	69
Ball and roller bearings	54	53	73	73
Knitting mills	54	52	71	67
Distilled liquor (except brandy)	54	47	71	73
Ceramic wall and floor tile	52	56	76	71
Commercial laundry equipment	51	53	63	65

W: The government withholds these data to avoid disclosing information about individual companies.

SOURCE: William A. Johnson et al., Competition in the Oil Industry (Washington: George Washington University, 1976) pp. 5, 6. Source cited there: 1967 and 1972 Census of Manufacturers: Concentration Ratios in Manufacturing (Washington: Department of Commerce, Bureau of the Census).

Appendix 6

LARGEST U.S. PRODUCERS OF CRUDE OIL
IN 1955

Rank and Company	Net Production[a] (000 barrels)	Percentage of Total	Cumulative Percentage
1 Standard Oil of N.J.	143,175	6.1	6.1
2 Texas	115,920	4.9	11.0
3 Shell Oil	95,220	4.1	15.1
4 Standard Oil of Calif.	86,595	3.7	18.8
5 Standard Oil of Indiana	83,145	3.5	22.3
6 Gulf Oil	79,695	3.4	25.7
7 Socony Mobil	79,350	3.4	29.1
8 Continental Oil	46,230	2.0	31.1
9 Phillips Petroleum	43,125	1.8	32.9
10 Sinclair[b]	41,055	1.7	34.6
11 Sun Oil	35,190	1.5	36.1
12 Union Oil of Calif.	35,190	1.5	37.6
13 Ohio Oil[b]	33,465	1.4	39.0
14 Cities Service	32,775	1.4	40.4
15 Tide Water Assoc. Oil[b]	30,705	1.3	41.7
16 Atlantic Refining	29,670	1.3	43.0
17 Amerada Petroleum	29,670	1.3	44.3
18 Sunray Mid Continent[b]	27,460	1.2	45.5
19 Skelly Oil[b]	23,460	1.0	46.5
20 Pure Oil[b]	23,115	1.0	47.5
Total: Top 20 companies	1,084,680		
Total: All U.S. companies	2,348,415		

[a]Barrels per day multiplied by 365. Barrels-per-day figures from Melvin de Chazea and Alfred E. Kahn, *Integration and Competition in the Petroleum Industry* (New Haven: Yale University Press, 1959), p. 30.

[b]Not on the 1970 list.

SOURCE: Data from *The Structure and Competitive Behavior of the Petroleum Industry—A Fact Sheet* (Washington: Federal Energy Administration, April 1976).

LARGEST U.S. PRODUCERS OF CRUDE OIL
IN 1970

1970 Rank and Company	1955 Rank	Production[a] (000 barrels)	Percentage of Total	Cumulative Percentage
1 Standard Oil of N.J.	1	376,614	10.7	10.7
2 Texaco	2	319,676	9.1	19.8
3 Gulf Oil	6	214,718	6.9	26.7
4 Shell Oil	3	204,085	5.8	30.5
5 Standard Oil of Calif.	4	177,331	5.0	37.5
6 Standard Oil of Indiana	5	159,838	4.5	42.0
7 Atlantic Richfield	16	151,503	4.3	46.3
8 Getty	N	134,456	3.8	50.1
9 Mobil	7	132,055	3.8	53.9
10 Union Oil of Calif.	12	95,902	2.8	56.7
11 Sun Oil	11	78,632	2.2	58.9
12 Marathon	N	63,820	1.8	60.7
13 Continental Oil	8	60,368	1.7	62.4
14 Phillips	9	47,677	1.4	63.8
15 Cities Service	14	45,001	1.3	65.1
16 Amerada Hess	17	30,879	0.9	66.0
17 Tenneco	N	29,576	0.8	66.8
18 Louisiana Land & Explor.	N	22,617	0.6	67.4
19 Superior Oil	N	18,607	0.5	67.9
20 Standard Oil of Ohio	N	10,497	0.3	68.2
Total: Top 20 companies		2,398,900		
Total: All U.S. companies[b]		3,517,450		

[a]Based on average daily production of crude oil as reported in annual reports and Moody's reports. In all cases, it was not possible to separate gross and net production.

[b]Bureau of the Mines, *Minerals Yearbook,* Washington D.C.; Government Printing Office, 1972, p. 817.

SOURCE: Data from *The Structure and Competitive Behavior of the Petroleum Industry—A Fact Sheet* (Washington: Federal Energy Administration, April 1976).

Appendix 8

Case	Entry	Number/Strength of Competitors	Reasons for Success	Defendant's Market Share
Major Cases Finding that Defendant Violated Section 2				
Alcoa (1945)	None	No domestic competition	Patent monopoly, cartels, preemption of raw materials, expanded capacity faster than sales and was content with a low rate of return.	80-90% of industry for 25 years
United Shoe (1953)	Only one significant entrant	No significant competitors	Merger, acquisition, discriminatory 10-year leases	85% of market for 40 years
American Tobacco (1946)	None in 8 years	Three large and several small competitors	Conspiracy to fix prices and exclude competitors	Three competitors shared 75% of cigarette market for 40 years, declining slowly
Grinnell (1966)	None effective	No significant competitors	Mergers and agreements not to compete	87-91% of market
Major Cases Finding that Defendant Did Not Violate Section 2				
du Pont (1956)	Substantial	Many in flexible wrappings; only 2 in cellophane	Competitive achievement and willingness to take risks	75% of cellophane, 20% of flexible wrappings
Hughes Tool (1954)	Some successful entrants	Four significant competitors	Best product and excellent service	75% of roller bit industry and stable for 20 years

SOURCE: Edward J. Mitchell, editor, *Vertical Integration in the Oil Industry* (Washington: American Enterprise Institute, © 1976), p. 36. Source cited there: Cravath, Swaine & Moore, *Pretrial Brief for International Business Machines*, submitted to the U.S. District Court, Southern District of New York, 15 January 1975, pp. 4, 5.

DOMESTIC SELF-SUFFICIENCY OF
LEADING REFINERS IN 1969
(FTC Estimates)

Company	Self-Sufficiency (percentage, runs to stills)
Standard of N.J.	87.4
Standard of Indiana	50.5[a]
Texaco	81.0[b]
Shell	62.1
Socal	68.8[a]
Mobil	42.2[c]
Gulf	87.6[a, d]
ARCO	64.9
Sun	46.7[e]
Union	64.3[a]
Standard of Ohio	6.7[a]
Phillips	51.8[a]
Ashland[f]	12.6
Continental	64.0
Cities Service	49.9
Getty[g]	137.2[d]
Marathon	88.1

[a]Other liquids included in crude production.
[b]Estimated.
[c]Other liquids included in refinery runs.
[d]Excludes crude processed for company's account.
[e]Crude production includes Canada.
[f]Twelve months to 30 September 1969.
[g]Includes subsidiaries.

SOURCE: Edward J. Mitchell, editor, Vertical Integration in the Oil Industry (Washington: American Enterprise Institute, © 1976), p. 66. Source cited there: FTC, "Preliminary Staff Report," July 1973, p. 20.

Appendix 10

LARGEST HOLDERS OF KNOWN U.S. COAL RESERVES, 1970

Rank and Company	Reserves (millon tons)	Location	Type	Data Source
1 Burlington Northern	11,000	W	l-s	Fo
2 Union Pacific	10,000	W	50% l-s	Fo
3 Kennecott Copper	8,680	(total)	mostly st	10K
(Peabody Coal)	3,800	M	bit	DE
	2,300	W	l-s	10K
4 Continental Oil	7,731	(total)	74% st	10K
(Consolidation Coal)	964	M		DE
	3,400	W	l-s	BW
5 Exxon	7,000	(total)		Fo
(Monterey Coal)	3,250	M	bit	DE
	3,000	W		BW
6 American Metal Climax	4,000	E,M,W	50% l-s	Fo
(Ayrshire Coal)				
7 Occidental Petroleum	3,300	(total)	28% l-s	10K
(Island Creek Coal)	1,163	M		DE
8 United States Steel	3,000	mostly E	met	Mo
9 Gulf Oil	2,600	(total)	8% l-s	Fo
(Pittsburgh & Midway Coal)	278	M		DE
	2,000	W		BW
10 North American Coal	2,500	E,W	80% l-s	Fo
11 Reynolds Metals	2,100	W	95% l-s	BW,Fo
12 Bethlehem Steel	1,800			Fo
13 Pacific Power & Light	1,600	W	l-s	Fo
14 American Electric Power	1,500		h-s	Fo
15 Kerr-McGee	1,500		60% l-s	Fo
16 Eastern Gas & Fuel	1,468	E	33% l-s	Fo
17 Norfolk & Western RR	1,400		99% l-s	Fo
18 Utah Construction & Mining	1,300	W	94% l-s	Fo
19 Westmoreland Coal	1,200	(total)	88% l-s	Fo
	327	E		Mo
20 Pittston	1,000	E	l-s	Mo,Fo
Total	74,679			

KEY

Location			Type		
E	=	East	l-s	=	sulfur content 1% or less
M	=	Midwest	h-s	=	sulfur content over 1%
W	=	West	st	=	steam
			bit	=	bituminous
			met	=	metallurgical

Data Source
BW = *Business Week,* Dec. 11, 1979, pp. 92-94
DE = Defendant's Exhibit 61, U.S. vs. Eastern Div.; No. 67Cl632
Fo = *Forbes,* Nov. 15, 1972, pp. 32-46
Mo = *Moody's Industrial Manuals*
10K = Forms 10K submitted by company to Securities and Exchange Commission

SOURCE: *The Structure and Competitive Behavior of the Petroleum Industry—A Fact Sheet* (Washington: Federal Energy Administration, April 1976).

LARGEST COAL MINING COMPANIES
IN 1974

Rank and Company	Production (tons)
1 Peabody Coal (Kennecott Copper)	68,104,076
2 Consolidation Coal (Continental Oil)	51,753,933
3 Island Creek Coal (Occidental Petroleum)	20,848,017
4 Amax Coal (Amax Inc.)	19,948,871
5 Pittston	17,381,911
6 (United States Steel)	16,389,000
7 Arch Mineral (Ashland Oil)	13,878,539
8 Bethlehem Mines (Bethlehem Steel)	13,347,625
9 North American Coal	9,771,563[a]
10 Peter Kiewit Cons. Mining Div.	9,697,000
11 Old Ben Coal (Standard Oil of Ohio)	9,451,880
12 Eastern Associated Coal (Eastern Gas & Fuel)	7,697,893
13 Westmoreland Coal	7,580,575
14 Pittsburgh & Midway Coal Mining (Gulf Oil)	7,528,174
15 (Utah International)[b]	6,955,000
16 (General Dynamics)	6,950,073
17 Texas Utility Generating (Tex.)[c]	6,500,000
18 (American Electric Power)	6,378,631
19 Rochester & Pittsburgh Coal	4,608,792
20 A. T. Massey Coal Properties (St. Joe Minerals)	4,303,357

NOTE: Name in parentheses = parent company.

[a]Includes production of Decker Mine (6,786,000 tons in Montana). Operated by Peter Kiewit Construction but jointly owned by Kiewit & Sons and Pacific Power & Light.

[b]In addition to coal, Utah International mines copper, iron ore, uranium, oil, and gas, in foreign countries as well as in the United States. It is also engaged in shipbuilding, development, and construction.

[c]Owned by Texas Utilities.

SOURCE: *The Structure and Competitive Behavior of the Petroleum Industry — A Fact Sheet* (Washington: Federal Energy Administration, April 1976).

Appendix 12

CONCENTRATION RATIOS IN
BITUMINOUS AND LIGNITE PRODUCTION

Year	Top 4 Firms	Top 8 Firms	Top 20 Firms
1955	17.8%	25.4%	39.5%
1960	21.4	30.5	44.5
1965	26.5	36.3	50.1
1970	30.7	41.2	56.5

SOURCE: *The Structure and Competitive Behavior of the Petroleum Industry — A Fact Sheet* (Washington: Federal Energy Administration, April 1976). Source cited there: *Keystone Coal Industry Manual: U.S. Coal Production by Company,* 1955, 1960, 1965, and 1970 editions.

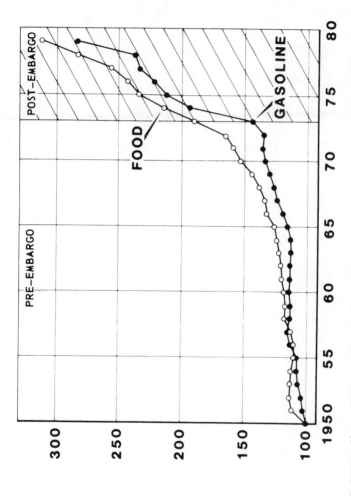

PRICE PATTERNS OF FOOD AND GASOLINE

PRE-EMBARGO

POST-EMBARGO

FOOD

GASOLINE

SOURCE: H. A. Merklein and W. R. Murchison, Jr., *Those Gasoline Lines and How They Got There* (Dallas, Texas: The Fisher Institute, 1980), p. 75. Sources cited there: gasoline prices, 1950–77, *Basic Petroleum Data Book* (Washington: American Petroleum Institute); 1978–79, *Monthly Energy Review* (Washington: Department of Energy, July 1979); food prices, calculated from *Handbook of Labor Statistics* (Washington: Department of Labor).

Appendix 14

ENERGY CONSUMPTION PER GNP DOLLAR

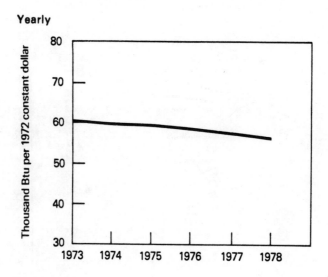

SOURCE: *Monthly Energy Review—Executive Summary* (Washington: Department of Energy, Energy Information Administration), September 1979, p. 17.

U.S. PASSENGER CAR EFFICIENCY

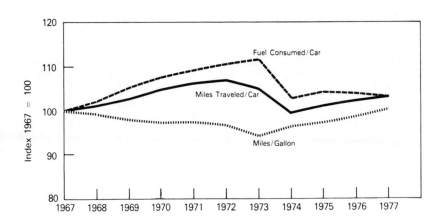

SOURCE: *Monthly Energy Review—Executive Summary* (Washington: Department of Energy, Energy Information Administration), September 1979, p. 19. Source cited there: "Highway Statistics," Table VM-1 (Washington: Department of Transportation, Federal Highway Administration, Federal Highway Statistics Division).

Appendix 16

RETAIL REGULAR GASOLINE PRICES
IN SELECTED DEVELOPED COUNTRIES
(U.S. Cents per U.S. Gallon)

COUNTRY AND DATE	PRICE INCL. TAX	PRICE EXCL. TAX	FEDERAL TAX	PERCENTAGE OF INCREASE OCTOBER 1973 TO MAY/JUNE 1979 Price Incl. Tax	Price Excl. Tax	Federal Tax
United States						
1973 October	40	28	12			
1975 January	53	41	12			
1976 January	58	46	12			
1977 January	60	48	12			
1978 January	62	50	12			
1979 May	81	69	12	102.5	146.4	0
France[a]						
1973 October	110	35	75			
1975 January	150	65	85			
1976 January	156	68	88			
1977 January	185	73	112			
1978 January	194	76	118			
1979 June	235	79	156	113.6	125.7	108
Italy[a]						
1973 October	79	19	60			
1975 January	129	41	88			
1976 January	135	46	89			
1977 January	216	61	155			
1978 January	216	60	156			
1979 June	216	60	156	173.4	215.8	160
United Kingdom						
1973 October	60	22	38			
1975 January	118	72	46			
1976 January	125	62	63			
1977 January	132	67	65			
1978 January	126	61	65			
1979 June	182	105	77	203.3	377.3	102.6
West Germany						
1973 October	144	48	104			
1975 January	166	58	108			
1976 January	182	74	108			
1977 January	186	78	108			
1978 January	181	73	108			
1979 June	200	90	110	38.9	125.0	5.8

[a]Government price ceiling in effect.

NOTE: Prices converted at March 1, 1979, exchange rates.

SOURCE: *International Energy Statistical Review* (Washington: Central Intelligence Agency, National Foreign Assessment Center), September 19, 1979, p. 20.

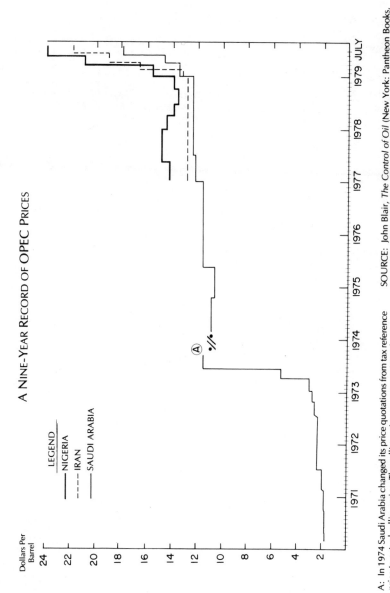

A NINE-YEAR RECORD OF OPEC PRICES

A: In 1974 Saudi Arabia changed its price quotations from tax reference price to actual selling price. The selling price was approximately 93 per cent of the tax reference price to 1974. Pre-1974 prices on this graph are tax reference prices.

SOURCE: John Blair, *The Control of Oil* (New York: Pantheon Books, 1976), p. 263, and *Monthly Petroleum Review* (New York: Merrill Lynch, Pierce, Fenner & Smith), October 1979, p. 22.

111

Appendix 18

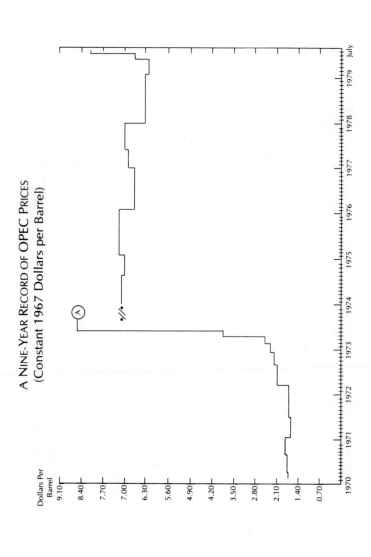

A NINE-YEAR RECORD OF OPEC PRICES
(Constant 1967 Dollars per Barrel)

A: In 1974 Saudi Arabia changed its price quotations from tax reference price to actual selling price. The selling price was approximately 93 per cent of the tax reference price to 1974. Pre-1974 prices on this graph are tax reference prices.

SOURCE: John Blair, *The Control of Oil* (New York: Pantheon Books, 1976), p. 263, and *Monthly Petroleum Review* (New York: Merrill Lynch, Pierce, Fenner & Smith), October 1979, p. 22.

CRUDE OIL WELLHEAD PRICES[a] 1948-79
(Current and Constant[b] 1967 Dollars per Barrel)

YEAR	DOMESTIC AVERAGE		YEAR	DOMESTIC AVERAGE	
	Current	Constant		Current	Constant
1948	2.60	3.14	1964	2.88	3.04
1949	2.54	3.23	1965	2.86	2.96
1950	2.51	3.07	1966	2.88	2.89
1951	2.53	2.78	1967	2.92	2.92
1952	2.53	2.86	1968	2.94	2.85
1953	2.68	3.07	1969	3.09	2.89
1954	2.78	3.17	1970	3.18	2.89
1955	2.77	3.15	1971	3.39	2.97
1956	2.79	3.08	1972	3.39	2.85
1957	3.09	3.31	1973	3.89	2.88
1958	3.01	3.18	1974	6.87	4.29
1959	2.90	3.06	1975	7.67	4.38
1960	2.88	3.03	1976	8.19	4.47
1961	2.89	3.06	1977	8.57	4.42
1962	2.90	3.06	1978	8.96	4.29
1963	2.89	3.06	1979	9.76	4.34

[a]Crude oil wellhead prices were derived by dividing the sum of the value of all first purchasers' purchases by the total volume of all first purchasers' purchases.

[b]The Producer Price Index for All Commodities (1967 = 100) was used to derive the Constant Prices.

SOURCE: Data from the U.S. Department of Energy, Energy Information Administration, *Annual Report to Congress 1978,* Volume 2 (of 3), 1979, except where otherwise noted.

Appendix 20

CRUDE OIL WELLHEAD PRICES[a] 1974-79
(Current and Constant[b] 1967 Dollars per Barrel)

YEAR	LOWER TIER		UPPER TIER		ALASKAN NORTH SLOPE[c]		STRIPPER OIL		DOMESTIC AVERAGE	
	Current	Constant	Current	Constant	Current	Constant	Current	Constant	Current	Constant
1974	5.03	3.14	10.13	6.33	...	...	...	...	6.87	4.29
1975	5.03	2.87	12.03	6.87	...	...	...	...	7.67	4.38
1976	5.13	2.80	11.71	6.39	...	...	12.16	8.19	8.19	4.47
1977	5.19	2.68	10.72	5.53	6.35	3.27	13.59	8.57	8.57	4.42
1978[d]	5.44	2.60	10.23	4.89	5.22	2.49	13.94	8.96	8.96	4.29
1979[e]	5.79	2.57	12.80	5.69	6.44	2.86	14.79	6.57	9.76	4.34

[a]Crude oil wellhead prices for each category and for the domestic average were derived by dividing the sum of the value of all first purchasers' purchases by the total volume of all first purchasers' purchases.

[b]The Producer Price Index for All Commodities (1967 = 100) was used to derive the Constant Prices.

[c]Alaskan North Slope crude oil prices were reported as Upper Tier prior to July 1977. These figures were obtained from Kenneth J. Arrow and Joseph P. Kalt, Petroleum Price Regulation: Should We Decontrol? (American Enterprise Institute, 1979).

[d]Average based on January through November reported data.

[e]The 1979 figures are an average of the prices for the various categories and indexes for the first four months of 1979. Source: U.S. Department of Labor Statistics, Monthly Energy Review, September 1979.

NOTE: Prior to February 1976 Lower Tier crude oil was called Old Oil and Upper Tier crude oil was called New Oil. Alaskan North Slope crude oil is included in Upper Tier and Domestic Average for all years shown.

SOURCE: Data from the U.S. Department of Energy, Energy Information Administration, Annual Report to Congress 1978, Volume 2 (of 3), 1979, except where otherwise noted.

Appendix 21

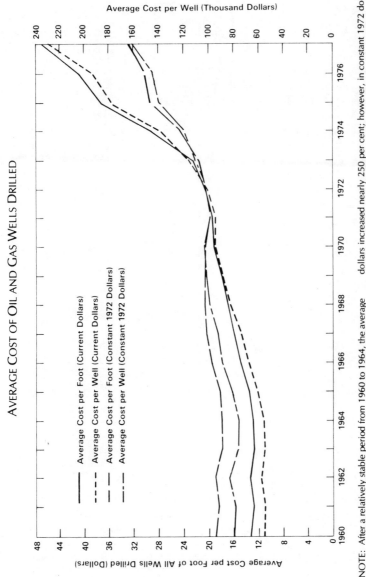

AVERAGE COST OF OIL AND GAS WELLS DRILLED

Average Cost per Well (Thousand Dollars)

Average Cost per Foot of All Wells Drilled (Dollars)

— Average Cost per Foot (Current Dollars)
- - - Average Cost per Well (Current Dollars)
— – Average Cost per Foot (Constant 1972 Dollars)
— · — Average Cost per Well (Constant 1972 Dollars)

NOTE: After a relatively stable period from 1960 to 1964, the average cost per foot for wells drilled increased at an average annual rate of 11.6 per cent from 1965 through 1977. In constant 1972 dollars, the average annual increase during this period was 5.3 per cent.
From 1960 through 1977 the average cost per foot drilled in current dollars increased nearly 250 per cent; however, in constant 1972 dollars, the increase was only 83 per cent.

SOURCE: Annual Report to Congress 1978, Volume 2: Data (Washington: Department of Energy, Energy Information Administration, 1979).

115

Appendix 22

CATEGORIES OF DOMESTIC CRUDE OIL PRODUCTION
(Percentage)

Year	Lower Tier	Upper Tier	Stripper Oil	Alaskan North Slope[a]	Naval Petroleum Reserves[b]
1976 average	54.4	31.6	14.0	....	...
1977 average	45.9	36.1	13.3	4.1	0.5
1978 average	37.5	34.4	14.0	13.0	1.1
1979 average (first 5 months)	34.9	34.7	15.0	14.3	1.2

[a]Alaskan North Slope crude oil prices were reported as Upper Tier prior to July 1977.

[b]The federally owned Naval Petroleum Reserves are exempt from price controls but were reported as Upper Tier prior to July 1977.

SOURCE: Calculated from U.S. Department of Energy, Monthly Energy Review, as cited by Kenneth J. Arrow and Joseph P. Kalt in Petroleum Price Regulation: Should We Decontrol? (Washington: American Enterprise Institute, © 1979).

TOTAL FOOTAGE DRILLED
GAS, OIL, AND DRY HOLES
(Millions of Feet)

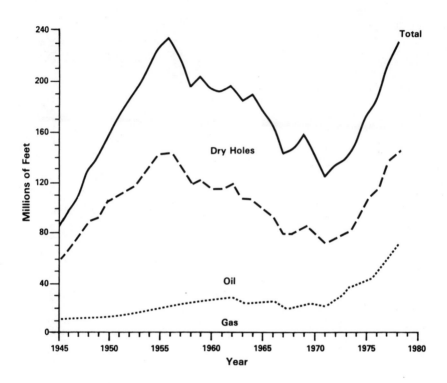

NOTE: Dry-holes data can be derived by subtracting oil and gas footage from total footage. Oil footage drilled can be derived by subtracting gas footage from total oil and gas footage. Total footage excludes service wells and stratigraphic and core tests.

SOURCE: *Historical Review of Domestic Oil and Gas Exploratory Activity* (Washington: Department of Energy, Energy Information Administration), p. 5. Sources cited there: data for 1945 to 1965 from *World Oil* magazine; data for 1966 to 1978 from *Quarterly Review of Drilling Statistics for the United States, Annual Summaries.*

Appendix 24

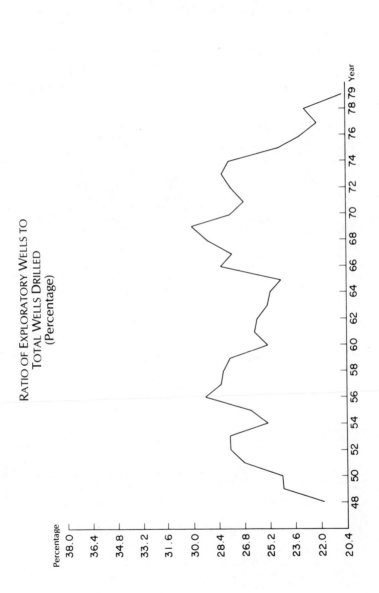

RATIO OF EXPLORATORY WELLS TO
TOTAL WELLS DRILLED
(Percentage)

SOURCES: Data from *Annual Report to Congress, 1978* (Washington: Department of Energy, Energy Information Administration), vol. 2, pp. 27-29, and, for 1979, from *Monthly Petroleum Review* (New York: Merrill Lynch, Pierce, Fenner & Smith), October 1979, p. 43. The 1979 figure is the average of the January through August ratios.

AVERAGE DEPTH OF NEW FIELD WILDCAT WELLS
(Thousands of Feet)

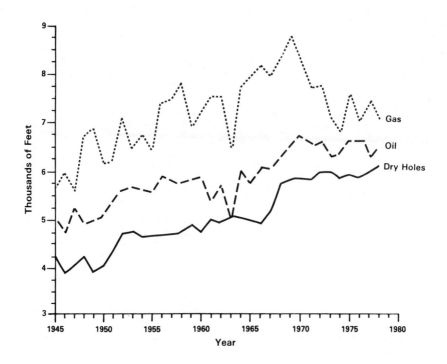

NOTE: Average depth is calculated by dividing footage by number of wells.

SOURCE: *Historical Review of Domestic Oil and Gas Exploratory Activity* (Washington: Department of Energy, Energy Information Administration), p. 17. Source cited there: *Quarterly Review of Drilling Statistics for the United States.*

Appendix 26

HYDROCARBON RESERVES ADDED
PER FOOT DRILLED
(Barrels of Oil Equivalent per Foot)

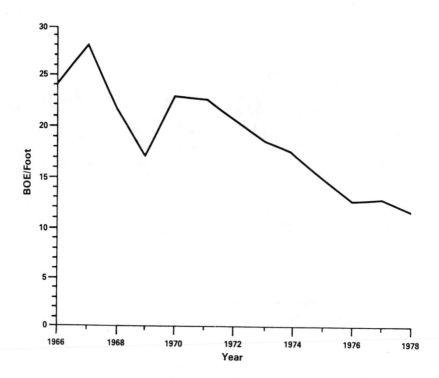

NOTE: Prior to 1966, natural gas reserves data did not distinguish between revisions and extensions. Revision and Prudhoe Bay addition are excluded from reserve additions. Natural gas liquids and natural gas are converted to barrels of oil equivalent using the following heat contents: crude oil contains 5.8 million BTUs per barrel; natural gas liquids contain 4.023 million BTUs per barrel; natural gas contains 1.032 million BTUs per thousand cubic feet. Total footage drilled excludes service wells and stratigraphic and core tests.

SOURCE: *Historical Review of Domestic Oil and Gas Exploratory Activity* (Washington: Department of Energy, Energy Information Administration), p. 23. Source cited there: *Reserves of Crude Oil, Natural Gas Liquids and Natural Gas in the United States and Canada as of December 31, 1978*, pp. 24, 116, 117, 120.

ACTIVE ROTARY RIG COUNT
(Annual Average Activity)

SOURCE: *Historical Review of Domestic Oil and Gas Exploratory Activity* (Washington: Department of Energy, Energy Information Administration), p. 15. Source cited there: Hughes Tool Company.

121

Appendix 28

SEISMIC CREW MONTHS WORKED

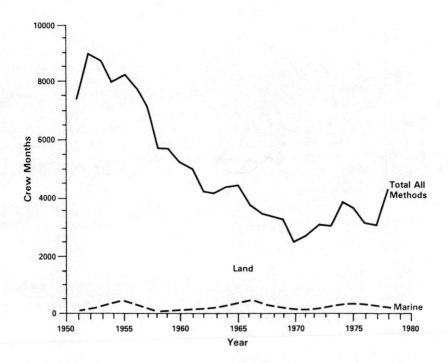

NOTE: Land crew months may be derived by subtracting Marine data from Total All Methods.

SOURCE: *Historical Review of Domestic Oil and Gas Exploratory Activity* (Washington: Department of Energy, Energy Information Administration), p. 21. Source cited there: *Annual Geophysicist,* Society of Exploration Geophysicists.

122

RESERVES TO PRODUCTION RATIO

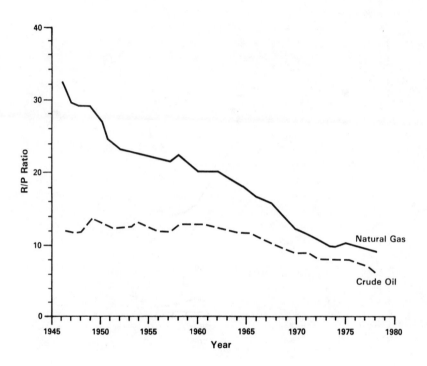

NOTE: Prudhoe Bay was excluded since production from this field did not begin until 1977. The inclusion of the Prudhoe Bay Field's 9.6 billion barrels and 26 Tcf of proved reserves in 1970 changes the R/P ratio as of 1/1/71, for example, from 9.030 to 11.978 for crude oil and from 11.992 to 13.170 for natural gas. Reserves to production ratio is calculated by dividing reserve balance as of the end of the year by that year's reported production. Natural gas reserves and production include both non-associated and associated-dissolved gas. Prior to 1966, natural gas reserve addition data did not distinguish between non-associated and associated-dissolved gas or between new field discoveries and new reservoir discoveries in old fields.

SOURCE: *Historical Review of Domestic Oil and Gas Exploratory Activity* (Washington: Department of Energy, Energy Information Administration), p. 51. Source cited there: *Reserves of Crude Oil, Natural Gas Liquids and Natural Gas in the United States and Canada as of December 31, 1978,* pp. 24, 116.

Appendix 30

PROVED CRUDE OIL RESERVES
(Billions of Barrels)

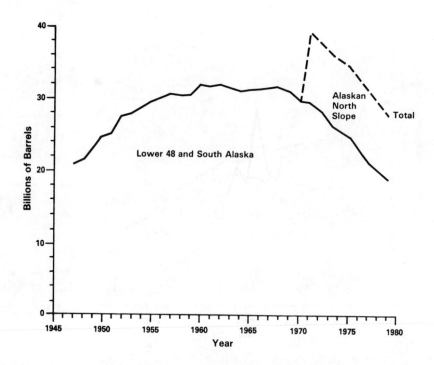

NOTE: Proved crude oil reserves as of the beginning of the year were used. Alaskan North Slope (Prudhoe Bay) can be derived by subtracting Lower 48 and South Alaska from Total.

SOURCE: *Historical Review of Domestic Oil and Gas Exploratory Activity* (Washington: Department of Energy, Energy Information Administration), p. 53. Source cited there: *Reserves of Crude Oil, Natural Gas Liquids and Natural Gas in the United States and Canada as of December 31,* pp. 24, 27.

CRUDE OIL RESERVE ADDITIONS VERSUS PRODUCTION
(Billions of Barrels)

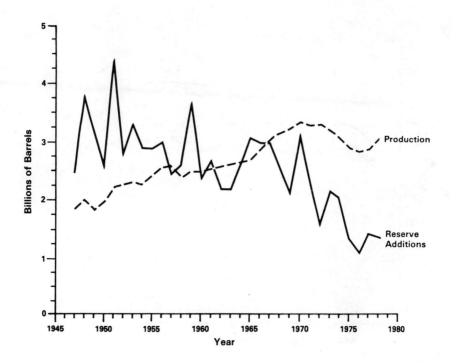

SOURCE: *Historical Review of Domestic Oil and Gas Exploratory Activity* (Washington: Department of Energy, Energy Information Administration), p. 65. Source cited there: *Reserves of Crude Oil, Natural Gas Liquids and Natural Gas in the United States and Canada as of December 31, 1978,* p. 111.

Appendix 32

CRUDE OIL AND NATURAL GAS LIQUIDS PRODUCTION
(Millions of Barrels per Day)

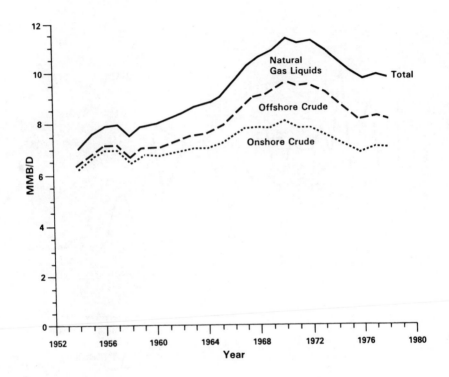

NOTE: The U.S. Geological Survey has estimated offshore production based on Bureau of Mines total production and data from state commissions. Offshore production includes both state and federal OCS. Natural Gas Liquids result from gas processing at natural gas processing plants.

SOURCE: *Historical Review of Domestic Oil and Gas Exploratory Activity* (Washington: Department of Energy, Energy Information Administration), p. 69. Sources cited there: *Annual Petroleum Statements*, U.S. Bureau of Mines; U.S. Geological Survey provided offshore crude oil production data for 1945 to 1971, and Bureau of Mines provided it for 1972 to 1976.

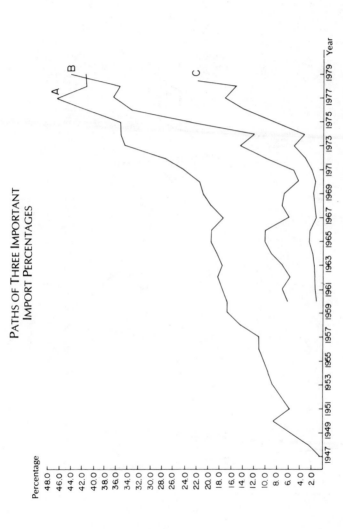

PATHS OF THREE IMPORTANT
IMPORT PERCENTAGES

A: Net oil imports as a percentage of domestic demand for refined products.
B: Imports from the Arab members of OPEC as a percentage of all imports.
C: Imports from the Arab members of OPEC as a percentage of domestic demand for refined products.

SOURCES: Data from *Petroleum Statement, Annual and December 1977*, and *Annual Report to Congress, 1978* (Washington: Department of Energy, Energy Information Administration) and from *Monthly Petroleum Review* (New York: Merrill Lynch, Pierce, Fenner & Smith), December 1979.

Appendix 34

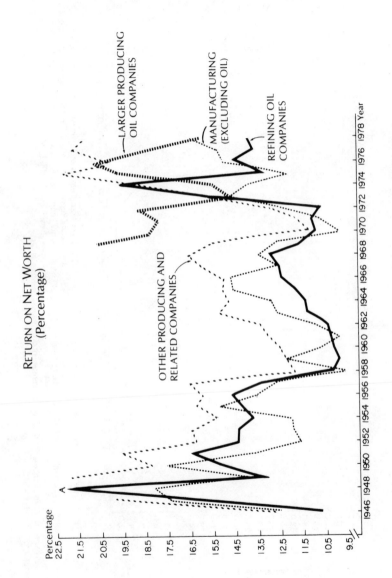

RETURN ON NET WORTH
(Percentage)

LARGER PRODUCING
OIL COMPANIES

MANUFACTURING
(EXCLUDING OIL)

REFINING OIL
COMPANIES

OTHER PRODUCING AND
RELATED COMPANIES

Percentage
22.5
21.5
20.5
19.5
18.5
17.5
16.5
15.5
14.5
13.5
12.5
11.5
10.5
9.5

1946 1948 1950 1952 1954 1956 1958 1960 1962 1964 1966 1968 1970 1972 1974 1976 1978 Year

A: For 1948 the return for Other Producing and Related Companies was 35.9 per cent, which is off the graph.

SOURCE: Data from *Monthly Economic Letter* (New York: Citibank N.A.), April 24, 1979.

CHANGE IN NET INCOME OF
LEADING OIL COMPANIES,
THIRD QUARTER 1973-1975

Company	1973-1975	1974-1975
Amerada Hess	(34.0)%	(24.3)%
American Petrofina	81.5	(47.6)
Ashland	59.0	12.4
Atlantic Richfield	64.2	(31.8)
Cities Service	52.9	(7.9)
Clark	(66.3)	140.8
Continental	52.4	(38.2)
Diamond Shamrock	157.7	26.5
Exxon	(13.8)	(31.3)
Getty	162.9	2.4
Gulf	(16.7)	(36.4)
Kerr-McGee	147.0	(2.7)
Marathon	(16.5)	(31.2)
Mobil	. . . . .	(16.8)
Murphy	(30.9)	(46.7)
Occidental	34.2	(60.1)
Pennzoil	44.9	(3.6)
Phillips	34.1	(36.0)
Shell	91.1	(26.0)
Standard Oil of Calif.	(15.5)	(32.7)
Standard Oil of Indiana	44.6	(28.2)
Standard Oil of Ohio	81.7	(19.1)
Sun	(76.0)	(30.2)
Texaco	(24.3)	(37.9
Union	62.1	2.9
Average	(3.1)%	(28.4)%

NOTE: Numbers in parentheses = percentage *decrease*.
SOURCE: Sherman Clark, *Oil Industry Earnings 1950-1975* (Claremont, Calif.: Claremont Men's College, 1976).

Appendix 36

AN INTEGRATED MAJOR'S EXPLORATION EXPENDITURES
IN RELATION TO ITS PROFITS

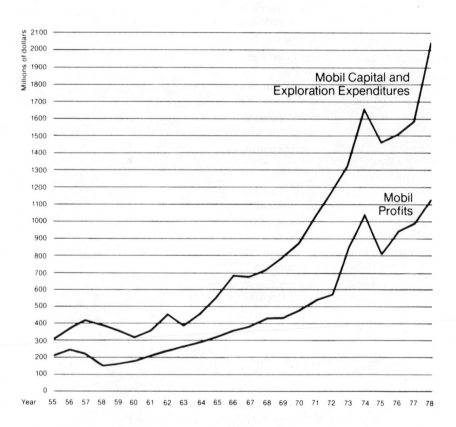

SOURCE: This graph appeared as part of a Mobil ad in the *New York Times,* April 26, 1979 (© 1979 Mobil
Corporation).

130

Bibliography

Arrow, Kenneth J., and Kalt, Joseph P. *Petroleum Price Regulation: Should We Decontrol?* Washington: American Enterprise Institute, 1979.

Blair, John M. *The Control of Oil.* New York: Pantheon Books, 1976.

Clark, Sherman. *Oil Industry Earnings 1950-1975.* Claremont, Calif.: Claremont Men's College, 1976.

de Chazea, Melvin G., and Kahn, Alfred. *Integration and Competition in the Petroleum Industry.* New Haven: Yale University Press, 1959.

Dobias, Richard S., *et al. Financial Analysis of a Group of Petroleum Companies — 1977.* New York: Chase Manhattan Bank, 1978.

Erickson, Edward W., and Waverman, Leonard, eds. *The Energy Question: An International Failure of Policy.* Volume 1: *The World.* Toronto: University of Toronto Press, 1974.

Erickson, Edward W., *et al.* "The Political Economy of Crude Oil Price Controls." *Natural Resources Journal,* October 1978.

Johnson, William A., *et al. Competition in the Oil Industry.* Washington: George Washington University, 1976.

Landsberg, Hans H., *et al. Energy: The Next Twenty Years.* Cambridge, Mass.: Ballinger Publishing Company, 1979.

McDonald, Stephen. *Petroleum Conservation in the United States: An Economic Analysis.* Baltimore: Johns Hopkins Press, 1971.

Mancke, Richard B. "Competition in the Oil Industry" in Edward J. Mitchell, ed., *Vertical Integration in the Oil Industry.* Washington: American Enterprise Institute, 1976.

————. *Performance of the Federal Energy Office.* Washington: American Enterprise Institute, 1975.

Medvin, Norman, *et al. The Energy Cartel: Big Oil vs. the Public Interest.* New York: Marine Engineers' Beneficial Association, 1975.

Mitchell, Edward J. *U.S. Energy Policy: A Primer.* Washington: American Enterprise Institute, 1974.

————, ed. *Vertical Integration in the Oil Industry.* Washington: American Enterprise Institute, 1976.

Murphy, Pamela. *Concentration Levels in the Production and Reserve Holdings of Crude Oil, Natural Gas, Coal and Uranium in the United States — 1955-1976.* Washington: American Petroleum Institute, 1977.

————. *U.S. Petroleum Market Volumes and Market Shares, 1950-1976 — Individual Company Data.* Washington: American Petroleum Institute, 1977.

Stobaugh, Robert, and Yergin, Daniel, eds. *Energy Future: Report of the Energy Project at the Harvard Business School.* New York: Random House, 1979.

U.S. Department of Energy, Energy Information Administration. *Annual Report to Congress, 1978.* Volume 2: *Data.* 1979.

_____. *The Determinants of Refinery Plant Size in the United States — An Analysis Memorandum.* 1978.

_____. *Effects of Oil Regulation on Prices and Quantities: A Qualitative Analysis.* 1979.

U.S. Department of the Treasury, Office of Economic Stabilization. *Historical Working Papers on the Economic Stabilization Program: August 15, 1971, to April 30, 1974, Part II.* 1974.

U.S. General Accounting Office. *The United States Refining Policy in a Changing World Oil Environment — Report to Congress.* 1979.

U.S. Office of Management and Budget. *Budget of the United States Government, Fiscal Year 1980.* 1979.

Ethics and Public Policy Reprints

Reprints are $1 each. Postpaid if payment accompanies order.
Orders of $10 or more, 10 per cent discount.

DATE DUE